CW00598092

ECONOMICS, SOCIETY AND VALUES

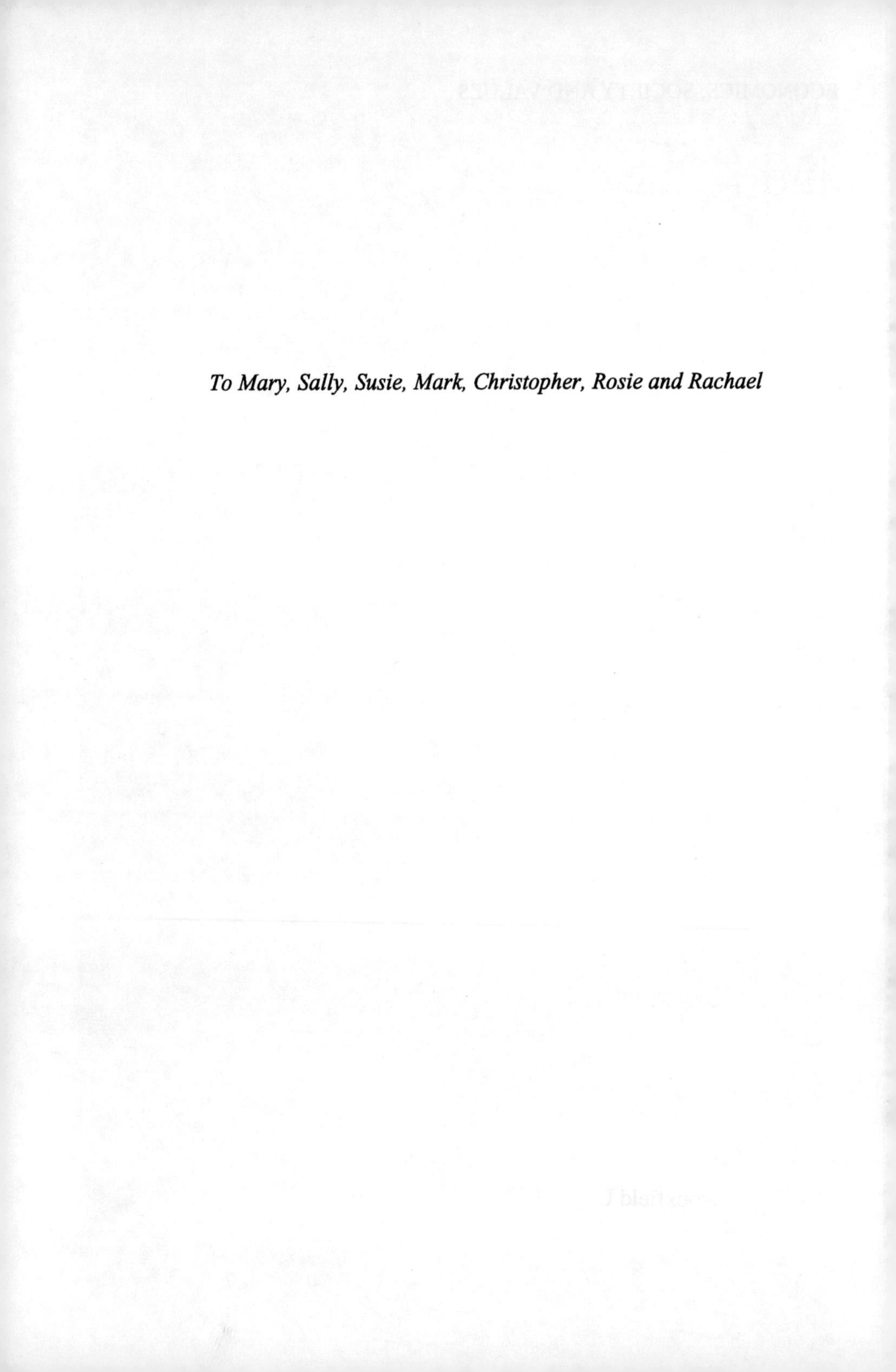

To Mary, Sally, Susie, Mark, Christopher, Rosie and Rachael

Economics, Society and Values

OWEN NANKIVELL

Avebury

Aldershot • Brookfield USA • Hong Kong • Singapore • Sydney

© Owen Nankivell 1995

All rights reserved. No part of this publication may be reproduced, stored in a retrieval system, or transmitted in any form or by any means, electronic, mechanical, photocopying, recording or otherwise without the prior permission of the publisher.

Published by
Avebury
Ashgate Publishing Limited
Gower House
Croft Road
Aldershot
Hants GU11 3HR
England

Ashgate Publishing Company
Old Post Road
Brookfield
Vermont 05036
USA

British Library Cataloguing in Publication Data

Nankivell, Owen
Economics, Society and Values
I. Title
330.01

ISBN 1 85628 866 8

Library of Congress Catalog Card Number: 95-80350

Printed and bound by Athenaeum Press, Ltd.,
Gateshead, Tyne & Wear.

Contents

Figures and tables

Acknowledgements

I owe much to many people in the writing of this book. Foremost, is my debt to my two colleagues and friends, my co-Directors of the Hinksey Centre, Raymond Clarke and Kenneth Wilson. The book is, above all, a product of that venture conceived thirteen years ago in a small restaurant in Kings Cross, London. I dare to hope that they will see in the book much of what we have tried to say and preach over these years. The support of the Centre's Reference Board has also been vital in encouraging me to persevere. I must also mention the many American colleagues and friends who have associated themselves with, and contributed much to, the Centre's work. My colleagues at St Olaf College, Northfield, Minnesota, have had an influence greater than I think they realize. I also dedicate the work to colleagues at the Vesper Society in Oakland, California; for the many years of productive and enjoyable collaboration. My appreciation, too, to Claude Gruson with whom I share so much in common. Finally, I thank Philip Klein of Penn State University, colleague and friend, who has always been there to remind me of the great American liberal tradition.

The physical task of getting a book to the point of publication is heavy and much of the load has fallen on my secretary and colleague, Linda Lear, to whom I am especially grateful. I also wish to thank Denis Loveridge for his painstaking editing of early drafts of the manuscript. The quality of the book, such as it is, owes much to his comments and suggestions.

Preface

This book is primarily for economists by an economist. But it is also written for my many colleagues in other disciplines; social sciences and value centred studies such as philosophy and religion. It has also been written for my friends in industry and, I hope, for the wider audience of those who want to make sense of economics; how it works, what role it plays in society and who are its masters.

That listing of the hoped for readership reflects my own vocational journey. I spent half my career in government service in the field of macro and micro economic forecasting in support of economic policy. I then made a career shift, not common in the British civil service tradition, by moving into industry. Five years as Chief Economist for a major manufacturing company, followed, since, by private sector consultancy work, has convinced me of the over-riding importance of a wider view of economic activity than that provided by the academic discipline of economics.

Sometimes I feel that I learnt more about economics in my first months with Lucas Industries than in all my years in Whitehall economic departments. What most struck me was the existence of a sense of identity of the company and the way it operated as a community. The external competitive environment was normally tough, the task of managing technological change hugely difficult, and the management atmosphere within the company was often competitive and adversarial. But the dominant impression was that, if the company was to succeed, success depended on its ability to combine together; to motivate the human resources that make up the community we call a company. At company level the conditions that have to be set to evoke that motivation include economic considerations such as pay levels. But there are many more conditions that relate to the quality of leadership or contribute to the sense of team spirit. They also relate to a sense of justice reflected in internal company practices and codes of treatment. They also reflect, indirectly, satisfaction or dissatisfaction with the wider social and political environment within which people were expected to work.

I also come from a background which makes it natural for me to take an interest in the ethical dimension of economic activity. The study programme at the Hinksey Centre, which I have been privileged to direct since 1982, in addition to my other activities, and which expresses a concern for value issues in economics, has underlined the importance of that ethical dimension. Economic activity (private and corporate), like any other activity, has to be judged from a moral standpoint. But the ethical dimension intrudes into a critique of economic activity in all sorts of other ways, which will be clear as the argument develops.

I intend to focus on the interaction between economics and society. I use the term economics in the widest possible sense; to cover economic activity generally, the theoretical structures put forward by economists to explain that economic activity and the, generally hidden, value assumptions underlying economic theory and practice. I bring in society because, arguably, economic activity is the principal arena for life fulfilment in western society today. Western societies are materialistic in the obvious sense of being rich, but also in the sense that they have opted to use economic activity, the processes of producing goods and services and the use made of the end results, to judge achievement - in personal terms and in community.

I appreciate that, anticipating the wide ranging analysis of economics that is to follow, I shall disturb many of my fellow economists whose view of the scope of economics is more narrow and more limited - on safer ground, some would say. But I make no apology for that. My view of economics is a high one. I think its role is all pervasive. It affects western lives profoundly. It has much to commend it for its concentration on solving the economic problem of scarcity and ensuring resources are allocated efficiently. The mass enjoyment of standards of living, beyond the wildest dreams of our ancestors, is a result of which we can be proud. That is what the industrial system has delivered. I am also glad that, amongst my colleagues, there are many that devote themselves to the study of the internal logic of that system. There is a pursuit of truth which has to be encouraged and applauded even though, in the case of economics, the direction of research and the reasons for adopting certain interpretations, appear to be uncomfortably prone to political fashion.

But it seems to me idle to hold that that is the limit of the interest in economics as a discipline. On the contrary, the more realistic perception of economics is of an activity inextricably locked into vast forces at work which track out a path that society at large has to follow. For instance, we ignore at our peril the repercussions of the continuing flow of new developments in science and technology on economic activity. I will argue later that the emergence of new technologies, and the way they are exploited, is decisive in changing the way economics is practised and the way society at large is organized. Again, the chances of history appear to have a crucial influence on the precise form of economic development. The way Britain emerged as an economic power in the late eighteenth century took place within a quite different set of parameters to those that have determined when and how Japan would develop. These forces ebb and flow with such energy that they defeat attempts to pursue a narrow interpretation.

The wider perspective of social change is also critical to a proper understanding of economics. Economics, as a social science, is about people. The forces that determine what people want, how they wish to relate to each other, what balance between personal and community values is preferred, are matters for philosophy and sociology rather than economics. But the way these issues are determined, and their consequences, are of immense importance to the way in which economic activity is perceived and practised. Society at large creates an overall constraint, sets parameters, within which economics has to work.

Since social is a wider and economics a more narrow perspective, it is reasonable to conclude, therefore, that economics is critically subject to wider forces and constraints. But that conclusion is belied by a different, generally held, perception; that economics is all-important and pervades our lives. Economics is put forward as a system of universal rules which has to be observed if society is to continue to have the benefit of the steady and increasing flow of goods and services which are vital to achieving so many of society's immediate goals. It is a system, too, that individuals have to conform to if they are to have a chance of a job and a guaranteed share in that wealth. These universal rules are also presented as allegedly empty of values - as science generally is invariably claimed to be - so there is no point in individuals or society claiming that the results of economic activity are not fair. Economics thus suffers from a time related dogmatism. Proponents of the system at any one time claim it is universally true and demands compliance. No matter that this claim is belied by even a cursory historical and geographical analysis.

My task will be to establish a wider context which will identify the complex relationships between economics and technology and their reaction with social development at large necessary for a proper understanding of the role of economics. By so doing, I hope to make a contribution to that necessary and vital task of all research - to provide a means, by better understanding, by which people can achieve greater life fulfilment. At the end of the day, economics, a social science involving people, who live and die within the context of that system, has to be judged against that criterion.

Will this argument then lead to a new formulation of economics based on more congenial precepts but less efficient in producing and allocating resources? Far from it. I will argue that it is only when an economic model has a shape and a value basis consistent with the norms and practices currently adopted in society at large, that the motivation to succeed in economics is fully harnessed. The challenge is to make that happen. The past shows that some societies succeed and some fail. The past also shows that within the constant ferment of ideas and of competing powers, the challenge is constantly being renewed. The moves towards greater unity in Europe present Britain with a contemporary challenge of substantial dimension.

Part I
THE DYNAMISM OF CHANGE

1 Introduction

Arthur Lewis, during his time as Professor of Economics at the University of Manchester, was said to have posed a standard question after every paper presented at his post-graduate seminar. 'What will this do to raise the world's standard of living?' A timely reminder that the discipline of economics is to be judged by its value to society.

It is easy to lose sight of a higher purpose in economic activity. Modern societies are hugely materialistic and contain sophisticated industrial structures of global proportions. The discipline of economics itself has now a vast corpus of technical knowledge. In its concern for the stuff of everyday life it addresses the predicament of all, even those as far apart as the poorest African peasant and the richest of western millionaires. Economic activity is increasingly, some would say, challenging the stability and long term viability of the planet's present resources.

Economics is about scarcity and the attempt to deal with that scarcity by way of choice and allocative rules; the essential economic problem. Society's materialistic wants persistently outstrip its capacity to supply goods and services. That capacity has increased over time to enable it to deliver more, but like the fate of Sisyphus, since it tries to satisfy the insatiable, the goal is never reached. Given this chronic inability to satisfy wants, the task of trying to resolve the economic problem has been constantly to improve efficiency. That is to say, to find ways of achieving more output with the same amount of factors of production; the constant quest for improvement in productivity of capital and/or labour. Efficiency is therefore a key word in economics. It underlines the unacceptability of waste since that implies an inefficient use of resources. It justifies the market clearing mechanisms of western market capitalism as a means of matching up supply and demand; a market which allows choice to be expressed and favours the efficient as against the inefficient in the meeting of the wants expressed by that choice.

But efficiency is a slippery term. We may have the best of goals but seek to meet them inefficiently. We can be ultra-efficient but devote our efforts towards unacceptable goals. In other words, it is not really possible to talk about economic efficiency except in relation to goals. This 'higher efficiency', as Professor Philip Klein puts it, 'refers to the ability of total resource allocation to reflect fully changing societal values, including notions of equity, security, compassion and freedom along with economists' usual concern with narrow efficiency'. [1] That is the duty of economics and it cannot, must not, set itself apart from social pressures to make choices based on prevailing values. 'There is a value floor to which economic analysis, among other kinds of analysis, must be related and without which it cannot be interpreted'. [2]

We will spend much time on this inter-relatedness between economics and value floors, to use Professor Klein's term, that underpins economics and society at large. That inter-relationship has been historical, in the sense that it can be observed in most past periods, and it is global, in the sense that it can be seen in different cultures existing in the world today. The form of the inter-relationship has not remained the same over time nor is it the same throughout the world today. This is because, as we will argue, the fundamental nature of economic activity is to be found in the fact of change. Driven by technology, the economic system is always on the move and consequently - such is its importance within the social framework - it seeks to carry all before it. The stakes then become high. They involve the nature of economic activity, the relationship of that activity to social institutions and the challenge to find the value base which best enables economic change to flourish.

Periods of dramatic economic change do not happen very often, but when they do, they open up great vistas and set society off in new breathtaking directions. The move from nomadic to settled agriculture was possibly one such time. The industrial revolution starting in the middle of the eighteenth century was clearly another. The twentieth century has been witnessing another such transformation which has not yet run its full force.

A closer examination, however, will show that the economy is always on the move. Even relatively quieter moments, such as the sixteenth and seventeenth centuries in Britain, contained important technological shifts with subsequent economic and social changes. Change is the common factor; the difference is that it may be relatively slow at times and relatively or dramatically fast at others.

The technological and economic developments that combined to create the period 1750 to 1850, conveniently known as the first industrial revolution, were clearly of a radical nature. New technologies led to new processes and new products which themselves led to radical changes in society at large; not least in changes in the values base against which these activities were justified. The evidence of the period is unequivocal in its documentation of these changes. Although the period since that revolution has been one of constant change and innovation, both economic and social, the speed of change does seem to have accelerated since the end of the second world war to give a new sense of revolution. This is particularly true of the past twenty years, which have seen the introduction of a whole range of new technologies, notably information technology. A process of innovation, of comparable proportions to the first industrial revolution, with consequential economic and social change, is taking place affecting the range of products on offer, the ways in which they are produced and transforming social institutions and lifestyle.

It would be very surprising if these current changes did not create social consequences at least as massive as those associated with the first industrial revolution, for economic and social change are always inter-related, each one affecting the other. It would be surprising, also, if there are not shifts, as occurred in the eighteenth and nineteenth centuries, in the value basis upon which society opts to justify both economic activity and its attitude to welfare and community relations.

We shall start by looking at past examples to identify the sequence of events, technological, economic, social and ideological, that occurred. A natural starting point will be the example of the industrial revolution of the eighteenth and nineteenth centuries, which should build up an idea of scale and inter-relatedness that occurs during periods of rapid change. The changes in technology were themselves many and varied, which poses a problem of finding the best means of focusing the argument. For ease of exposition, therefore, the consequences of technological change will be illustrated, in the main, by reference to the sole example, albeit a crucial one, of the introduction of steam power. The study of the consequential economic and social changes, however, will inevitably have to be more general and wide embracing.

Most will agree that starting about 1750 a process began in Britain by which a wave of new technologies (as part of the scientific revolution) transformed the nature and levels of economic activity. One of the principal features of this new economic activity - which brought in the era containing most of the features of modern life - was an ability to provide sustainable economic growth. As a result, it gave most people in most western societies the opportunity eventually to distance themselves from poverty to a degree and, for the growing number of the middle classes, an enjoyable fulfilling materialistic lifestyle. It also created an industrial structure based on large scale industry and an economic system which generated troublesome short term (cyclical) variations in the rate of economic growth. The new economic system also appeared - although this may be more debatable - to be subject to longer term cycles of some fifty years in length. Both the theory and the inferences to draw from the facts are to some extent disputed, although there seems to be a case for identifying major technological surges centred on changing from one source of power to another. But a feature of the new system was periods of alternating high and low rates of overall economic growth associated with rapid or slow rates of diffusion of technologies.

These technological and economic changes created consequential social effects. These included a displacement of labour enduring near intolerable working conditions and resulting ultimately in, amongst other consequences, the growth of unions and other working class movements. They also led to social legislation affecting such matters as safety at work, working hours, public health and education, electoral reform and the putting in place of what is now called the welfare state. The social pressures for reform emanated from the increased awareness of the need for social justice, as distinct from, and, perhaps, at the expense of the free play of individualism.

Schumpeter applied the term 'creative destruction' [3] to the process whereby new technologies replace old. It is a powerful term, aptly summarizing the driving forces at work at times of change. For beneath the seeming relative stability of overall growth rates can be found shifts at sectoral level where new technologies create entirely new industries and comprehensively wipe out old ones. But this concept of creative destruction can be more widely extended. For it is clear that the same powerful imperialistic thrust of the new to overcome the old applies to much more than technology. The same pressures can be seen to be at work affecting the form and location of industrial production, working habits and lifestyles, the form of major social institutions such as government, law and education, and, indeed, the value basis upon which social action is based. In all these areas creative destruction is continually taking place. The need for change, to destroy the old, is invariably a precondition to creating the conditions in which the advantage of the new can be fully exploited. This wider view provides the context for examining the changes that occurred during the first industrial revolution. This will be the task of Chapter Two.

We need then to return to the question of whether radical transformations of this nature are occurring again. The problem with the question, however, is that the whole of the period from the mid-nineteenth century, when the initial stimuli creating the first industrial revolution were fully exploited, to the present day, can be treated as a period of unabated change. Simply to start to list the technological changes that had taken place by the second world war makes the point; the use of oil, the motor car, air travel, new chemical and oil technology and plastics, radio and television, telephones.

However, there is a strong case for believing that the present wave of new technologies, beginning at the turn of the 1960s, once again confronts society with a period of radical change. Many have written about it arguing that the economy and society are again being transformed by new technologies. These new technologies are, in turn, leading to substantial shifts in economic structure and organization; they are also radically changing consumer patterns and encouraging new lifestyles.

What is emerging can conveniently be called the new service economy (since the switch from goods to service activity is a key feature). It will be examined in detail in Chapter Three in the same way as the first industrial revolution will be studied in Chapter Two. However, to anticipate for a moment a point to be made later, it will be noted that both industrial revolutions have served to establish unprecedented levels of per capita consumption. The growth in per capita consumption was a crucial and novel feature of the nineteenth century. It is a trend that has continued throughout the twentieth century and has now reached a level where consumption dominates economic activity and, because it is one of the main elements in satisfying personal goals, it provides a key link between economic activity and society.

Economists have a special interest in this growth in consumption. One of the most striking differences between, say, the eighteenth century and the twentieth century, as it would appear to someone from another planet, would be the increase in affluence of the general population. But the importance goes beyond the way economists analyze and interpret this field, i.e. the various theories of consumer behaviour, to raise a deeper philosophical point underlying consumerism now that most of western society is freed from the fear of poverty and basic survival. People now want rather than need goods and services and apparently want them without limit. The past two centuries have witnessed immense productivity gains and, in principle, society has a choice as to how it wishes to enjoy the benefits of these gains; in less work, for the same levels of consumption, or working at the same intensity and having increased consumption. It is true that leisure time has increased over the last two centuries. Daily hours of work have been reduced, the length of the working week fallen, longer holidays are taken and the age of retirement has also come down. However, the overwhelming proportion of the benefits of technology and productivity increases has been taken in increased supply of goods and services.

The economist's interest in consumption is relatively narrow and limited, for example, to a short term horizon such as forecasting the immediate future for fiscal and budgetary policy reasons so defined, it is an area where economists feel particularly confident, thanks to the last few decades' econometric research into consumer behaviour, and the working models derived from that work, relying mainly on explanatory variables such as relative prices and incomes. But when it comes to addressing the question of the underlying determinants of consumer behaviour, economists, if they are wise, will look to other disciplines like sociology and anthropology for answers. With the exception of some authors, few economists have attempted to widen the scope of models seeking to explain consumer spending, by introducing sociological factors. Ways of understanding consumption in this wider sense, why people consume (and the linked question of why people work), will be examined later in Chapters Three, Nine and Eleven.

Chapters Two and Three will track the consequences when, in the first instance, new technologies inaugurate a new period of rapid change. They will illustrate the extensive nature of the process of creative destruction and show that the process is to be found at work in areas outside the narrow field of product and process replacement. Powerful all-embracing pressures, following a similar logic, work to transform the economy, social institutions and the value sets against which they are justified. These processes are summarized in Chapter Four. The enlargement of the scope of the process of change involves a consideration of the inter-relationship between technology and economics, on the one hand, and the key institutions of society, on the other hand. Most of the latter are deeply entrenched in society, have evolved their own form and justify their activities by their own value sets. Many of them have an institutional form; they are real, they can be observed at work. But others, of equal importance, are less easy to pin down; are not substantial in the literal meaning of the word. The former category includes institutions such as the prevailing structure of industry, the legal or education sectors, the institutional church. The latter relate more to the realm of ideas and ideologies and include, for example, the heritage of the development of economic thought or the formulation and practice of value systems called in to justify respective actions (a task mostly hitherto pursued in religious circles). To make this distinction immediately highlights the difficulty of applying it precisely. The education system, for example, is more than teachers, schools or universities. It creates, sustains and defends its ideologies as strongly as anyone else.

As a group, these two areas can be called 'heavy structures' [4] which, whether they need to be modified or replaced or not, characteristically are difficult to change and have an influence which is often obscure yet pervasive. The interaction between the primary forces of change and these heavy structures must be examined and are the subject, in turn, of Chapters Five, Six, Seven and Eight.

What is the role of values and ideology in economic activity? The use of the word 'ideology' is perhaps a little frightening; certainly to many economists. However, it is used in order to make a simple point. Economics is one of the social sciences and therefore deals with people. It is not, and can never be, composed of an exact and neutral set of laws and principles like physics (if, indeed, physics itself comes under that category). Peoples' behaviour will only be understood if their motives are identified, and how, in given situations, they balance up competing values. These values issues are right at the heart of how individuals and community behave. In that sense economics is value-ridden - it is and it should be. To get at the heart of things in economics, especially economics, we need to look at the ways in which people organize together and balance the competing claims of individualism and community. It is, of course, an age-long tension - fully reflected in economic thought and action.

How important is the wider social framework in which economics has to be placed? It is clear that, as western economies emerged through the industrial revolution, the lines between economic and social perceptions became sharply drawn. Britain in the nineteenth century was a perfect example of this dialectic. The social consequences of the laissez-faire economic approach of the early nineteenth century are there to be seen still. But there is always a dialectic between economic and social demanding resolution. Chapter Nine will look especially at this relationship between economic activity and society.

Britain is not the only society that has experienced the changes linked to a modern capitalist structure. Many other countries have pursued the same line and most have differing cultural backgrounds. An important question, therefore, is whether these cultural differences matter. What, in fact, is the impact of cultural differences (which often reflect different ideologies) on economic performance? Some examples of cultural differences will be examined in Chapter Ten. It is an important area of study since the evidence of variety in culture and economic system, both over time and across country, needs to be placed in the balance as a counter-weight to those who argue for a converging value-free universal economic system. These days it is very difficult to restrain those who take for granted that the ideal world is where economic agents are able to compete freely across frontiers without hindrance: an international trading system sought for and defended by over-arching international organizations like the International Monetary Fund (IMF) and the General Agreement of Tariffs and Trade (GATT).

There is much value in the economic arguments for free trade, but a dominant theme of this study will be the need for a conditional acceptance of these claims. Despite the pressures to adopt a universal free trade model, countries are different both in the resources they have at their command, in the shape of their particular economic history, in the timing of the impact of change, and, above all, in their cultural patterns. These factors are likely to be crucial in explaining how individual economies seek to maximize both the welfare of their own community and that of the international community at large.

Taking this broad approach to an understanding of economic activity, i.e. that it has to embrace a wider social perspective and grapple with issues related to values and culture, will not go unchallenged, but it can be justified in the following way. Since economics is about people it must ultimately be judged in value terms. The use of the fruits of economic activity is to be evaluated by the extent to which it contributes to welfare and well-being.

Each society has to articulate its own view of welfare which it then must proceed to pursue as best it can. The creation of the view of welfare must be a positive act on the part of individuals or community. It would be bizarre, to say the least, to assume that the final use of goods and services was not based upon choice but on a distribution predetermined by an automatic set of rules. The question of whom the system is for is a prior criterion of judging performance. The economist's emphasis on the importance of consumer sovereignty is a useful contribution to this debate but it is unfortunately limited to the role of individuals as economic agents rather than as individuals seeking to pursue wider goals or wishing to act in community in pursuit of either economic or social goals. The process touches on value judgements all the time even if they are hidden or not recognized.

The debate over what use to make of the fruits of the system is not the only one where economic activity touches upon human activity. A survey of the past will show, as will a contemporary comparison between Britain and other cultures, that the choice as to how people relate together in the process of production, plays an important part in human fulfilment. Workplace conditions must satisfy certain basic ethical or moral criteria. Society now refuses to countenance young children working long hours or adults operating in working conditions with inadequate safety arrangements. If the debate were limited to these sorts of examples there would be danger that the issue would be seen simply as a conflict of objectives between optimal economic production at minimum cost and humane conditions of work. This is a common example of the conflict of values between economic activity and wider social considerations; that is to say, there may be reasons for 'softening' the optimum conditions for economic activity but a price has to be paid for it in terms of economic efficiency.

This latter limited debate needs to take place but it is more fruitful to address the proposition that, since performance at the workplace is essentially related to motivation, it is necessary to satisfy people's perceptions of what are right or good practices, if the best is to be drawn out of their performance. In some cases, the necessary conditions will be the provision of humane working conditions, but, more generally, they relate to other questions; the style of leadership, the freedom to perform or gain recognition as individuals, having a say in the overall direction of a business or, more widely, the fairness of the basic welfare and taxation structure. These conditions, because they contribute to life fulfilment, are crucial in determining the level of economic performance. The precise choice of which values matter in this respect appears to vary over time - the management model of the mercantilist period, if we can talk about such a thing, was substantially different to that now found in a modern business. But the choice also appears to vary between cultures; between, say, Japan and Britain or between Germany and the United States of America. This set of issues lies at the heart of much of the debate about the form and direction that the current move towards greater European union should take. That debate represents a great practical challenge to Britain by forcing it to decide on its 'Europeanism', so much so that Chapter Twelve is devoted to applying the arguments of the book to the European debate.

The choice of the value sets upon which economic activity is to be based varies over time and with geography. It follows that it is important to expose the value sets upon which economic activity is based, in order to understand them better. We need to explore whether the best set of values is being adopted (judged, for example, by whether they are delivering optimum economic performance). Unless the existence of this value dimension is acknowledged, economic activity cannot be addressed in the way it should; as an arena contributing vitally to life fulfilment in ways stretching way beyond the narrow objective of earning money.

The argument does not, however, rest there. Society, whether or not it has found the appropriate value set upon which to base economic activity, has also to reconcile the tension that arises from the use of different value sets. It is very likely that other institutions in society are based upon value sets different from those adopted in the economic sphere. Social welfare institutions, for example, are not likely to endorse the same ethical judgement on the deserving and the undeserving poor which is the fashion in the current market capitalist economic model. Society is faced with the problem of how to enable such different value sets to co-exist. Moreover, it is clear that the boundaries between value sets are not fixed. From time to time the momentum of new insights into values in one area seems to want to spill over into other areas. The flow is two directional. In the Victorian era, for example, and again today, such is the militancy of the economic model that it seeks to export and apply its value sets into other parts of society. The managerial approaches of the economic system are being pressed on legal, education and health institutions. The virtues of the market, to minimize cost and to express choice, are finding their way into social care decisions; a business plan almost seems mandatory to deal with an elderly person stricken with dementia. There have been times when the flow has appeared to be in the other direction. When social attempts have been made to modify the way business operates by, say, advocating increased workers' rights or giving more attention to environmental factors.

As Chapter Eleven will show, this aggressive morality, prone to conquest other parts of society, presents a great challenge as to how society resolves such territorial disputes. But it is a challenge to be welcomed. Society gains nothing by hiding such questions under a cloak of pseudo value-neutral rules. Indeed, a sign of a healthy society is where these issues are openly recognized and debated in the widest possible terms by the widest possible constituency.

The fact of continuing change in the economy and in society will emerge unassailably in this study. The periods of history examined suggest that the forces of change are immensely powerful. They appear to follow a sequence of technology - economic - social - values. In the sense that there is a flow of this nature and in this direction, it is justifiable to talk of an imperialism; of pressures of change emanating from one area, pushing all resistance aside to allow the new forces to be fully exploited. The process may be acceptable, but this is manifestly not always the case. Society contains in it a constant battle where the imperialism of new ideas and new values is often contested. It may be contested because some object to the immediate impact of, say, economic change. Others object to the new ideology to which the forces of change appeal. Others may object because the value sets of some areas of society are being attacked and replaced by new and inappropriate ideologies.

When all the argument is done, it is hoped that a better understanding will have been reached of the way the process of technological and economic change occurs, particularly in the context of today's society. By so doing, economics can be set in its wider social context. Hopefully, too, the appropriate relationship between ideology and economics, and the relevance of culture in relation to economic performance will also have been established. The need to address and resolve all these inter-related questions is nowhere greater than in the context of the current debate about the future of the European Union. That debate dramatically focuses on what economics is all about and how to plan a future for society.

Notes

1. Klein, P.A. (1994), *Beyond Dissent: Essays in International Economics*, M.E. Sharpe, London, p. 108
2. Klein, P.A., ibid., p. 192
3. Schumpeter, J.A. (1965), *Capitalism, Socialism and Democracy*, Unwin University Books, Chapter 11
4. See Gruson, C. and Ladrière, P. (1992), *Ethique et Gouvernabilité - un projet européen*, Presses Universitaires de France, Paris, Chapter V

2 Technology and change in the nineteenth century

However much we would wish for sharp distinctions in time, for dates when something began or something else clearly ended, precision is impossible. Nor are there self-contained areas. It may be tempting to categorize events and to isolate trends by giving them labels such as economic causes, social consequences, the philosophical background, etc., but such words, economic, social, philosophical, are untrustworthy, indeed dangerous companions in describing the past.

The truth is that the object of our interest, which we might call the economy on the move, or more ambitiously, society on the move, is a focus that we can apply to almost any period of history. We can be sure any period will yield its own fascination. In looking at particular periods we will always find it difficult to know when to start, because we are intruding into a continuum. We will also be reluctant to settle for a *terminus ad quem* because the 'end' of our enquiry will already be the beginning of the next.

It is also true that when we examine the events that occur, and the consequences triggered by those events, even if we begin in the limited field of economics, we find that the scientific, technological, economic and social (including the interplay with social values) are always inextricably mixed together. Indeed, not only are they mixed together, but it is the very devil to separate them out and to determine which event is cause and which effect.

The prudent approach, therefore, and one which unhappily most economists are reluctant to adopt, is to embrace an eclectic view of interpreting economic history; to look for satisfactory explanations of what economics is all about by exploring the far wider canvas of social and economic change, rather than confining ourselves by the self-imposed limitations that surround the modern practice of economics.

As was pointed out in Chapter One, despite this over-riding sense of a continuum, there are some [1] who would argue that there are at least one or two periods in modern history where, for one reason or another, something so dramatic occurred, and the consequences were so far reaching, that we can safely examine each of them as a single phenomenon. Toffler [2] talks of the first industrial revolution, being the move from nomadic to settled agriculture, the second industrial revolution, being the period from 1750 onwards and generally known as the industrial revolution. He argues that we are now in the midst of a third revolution (the Third Wave, as he calls it).

There is no harm in creating these categories. The first makes sense although the event was so far back in time that it will not bother us any more. The second was dramatic and parts of it are the main subject of this chapter. Some of the third will be discussed in the next chapter. However, here is a good example where we need to be on our guard, lest we make too great a distinction in time and nature between what was happening in, say, the seventy five years from 1750 onwards and in this century so far. The types and scale of application of new technologies, and their impact on standards of living and the social framework, were vastly different. But there has been a momentum for change, in all manner of things, that makes it difficult to draw strong dividing lines.

What do we mean by an economy (society) on the move? A convenient way of categorizing is to envisage a process which originates with scientific discovery, leads on first to its technological exploitation, then to its economic application and followed, inevitably, by social consequences. We will find later that it is not that simple - if indeed it ever sounds simple - and the logical sequence may not always be in the direction suggested. Nevertheless, it is not a bad working model and it will provide insights into some important matters. This process is particularly helpful in understanding periods when the rate of change in economic growth has clearly accelerated.

Often, if not always, the primary cause of the acceleration is a fresh wave of inventions, leading to new technologies capable of being applied in the economic process. The sequence by which new technologies gain first, a foothold and then achieve dominance is neither simple nor speedy. Initially a new technology has to seek to gain recognition, by demonstrating that it can perform the same function at least as well as an existing technology. Once this foothold has been gained, the new technology is used to demonstrate that besides doing the traditional task equally well, it can also offer marginal applications which consolidate its superiority over the new technology - but still largely in the context of the perceived role and function of the old technology.

A third stage is more exciting because at this point in the application of the new technology, it is realised that in order to achieve its full potential a radically new approach to its exploitation is required. It is seen to be able to do many things in different ways to the old technology. This gives the new technology an ascendancy which the old technology cannot match and which dooms the old technology to extinction. Although the battle between old and new is virtually over at this stage there is often a need for a final development, usually requiring changes to the economic or social infrastructure, before the new technology can be fully exploited.

The process by which a new technology is introduced is thus often hard and long. It has to fight against the interests vested in the old technology and against the mind set dictated by the old technology. Its introduction is also hampered by the inevitable faults and limitations of the early applications of the technology. Above all it cannot be fully exploited until both the thought processes and the infrastructure have opened out to accommodate the full potential of the new technology.

This dynamic is an exciting process which unleashes powers of awesome proportions. The early part of the industrial revolution contains many examples and to consider them all would require a full economic history of the period. But our argument can be adequately carried forward by considering one important example; the invention, exploitation and application of the steam engine.

The first modern practical application of steam as a source of power was devised by Thomas Savery and presented in a model to William III at Hampton Court in 1698. It worked by a combination of atmospheric pressure and steam and was used to raise water from one level to another. By so doing it created the same conditions as water power, i.e. the water that had been raised could then fall by gravity to turn wheels. Savery envisaged its widespread use in industry but it became apparent that its ability to raise the level of water was severely limited in height (a hundred feet at most). However, it was very successful when less powerful tasks were required - in gardens, for instance. The Savery engine was technically limited in other respects too. It used heat inefficiently (and was therefore relatively costly), was slow and involved a considerable safety risk.

The Savery engine was followed by Newcomen's engine which was launched in 1705/6 and replaced Savery's for commercial uses. The engine worked on the basis of a piston action created by an oscillating rod raising and lowering one end by successive applications of steam and cold water in a cylinder to create a vacuum. Although the first engines were clumsy - in part because of the difficulty of making the pistons and cylinders with sufficient precision - improvements were steadily introduced. The condensation process was improved, the link and timing between the taps and the motion of the bar were made automatic and dangers reduced by the introduction of a safety valve.

The result was an engine with substantial technical and cost advantages over the earlier Savery engine. Commercial building and selling of Newcomen's engines began in 1711 and their use spread quickly for pumping water out of mines, feeding reservoirs and locks and supplying drinking water. The potential for using it as a driving engine was already there and in 1758 the matter was raised at the Royal Society, but it was not pursued any further since it was found simpler to use the engine to pump up water and allow gravity then to turn a wheel. The common use of the combined engine and water wheel had many disadvantages, the loss of heat and the consequent cost of fuel being the most important. But it did reign supreme during the first half of the eighteenth century.

The next major step forward was as a result of the work of James Watt. Watt took out his first patent in 1769 which he regarded as essentially a method of improving the fuel efficiency of the Newcomen engine, a model of which excited his curiosity in 1763/4 when he was asked to repair one of the engines. Effectively he addressed two problems; the excessive quantity of fuel used to raise the temperature in the cylinder afresh with each stroke and the inadequacy of the condensation process. Watt's solution was a condenser separate from the cylinder. This was the substantive technical advance; subsequent inventions were minor and designed to improve further the efficiency of the engine.

The two other significant developments surrounding the Watt steam engine were the advances in methods of manufacturing the engine and eventually the revolutionary transformation in use from a pump to a traction engine.

The development of the manufacturing process was long and costly and ran into the economic and financial hazards that are invariably associated with the practical application of inventions. Watt first sought the help of John Roebuck, a coalminer, whom Watt had advised on the use of pumps at his mine. Roebuck offered to finance the practical application of the invention in return for a share of the profits. The first machine was inaccurately constructed and still retained design faults as a result of which the development took much longer than expected. Meanwhile Roebuck ran into financial difficulties ending in bankruptcy in 1773.

The relationship with Roebuck brought Watts' work to the attention of Matthew Boulton. There had been abortive attempts of the three of them to work together - attempts that failed largely due to Roebuck's lack of foresight. The latter's difficulty gave Boulton the opportunity to take on Roebuck's partnership with Watt and the development work was transferred to Boulton's Soho factory in Birmingham. Boulton was already established as an innovative and profitable metal manufacturer and was able to provide Watt with exactly the resources, technical and financial, required to turn the invention into a cost effective product. The skilled workers at Soho soon ironed out the technical difficulties of the original Watt/Roebuck prototype engine so that it was made to work effectively. Successive patents gave Boulton/Watt the monopoly for over twenty five years, but the development process was long and costly, and required heavy financing from Boulton's other activities. It was not until 1787 that the firm began to reap its reward.

The next and greater step forward occurred in the early 1780s when Watt (assisted by one of the foremen at the Soho factory, William Murdock) formulated and patented the Sun and Planet (Rotary) Motion. This development revolutionized the application of Watt's engine making it applicable across the whole field of industry rather than limited to a pumping action.

Already the steam engine freed those industries requiring power from the restraint of needing to be close to water. From now on the only constraint was access to supplies of coal; hardly a problem anyway in view of the excellent system of waterways that already existed. Other factors determining choice of location could now be taken into account, closeness to supplies of raw materials, to their markets or to urban areas where their labour force was to be found.

In fact the northern regions, already best suited for supplies of water power, remained the industrial areas for other reasons, their closeness to coal and iron supplies and the climatic advantages for textiles. Nonetheless, even before the invention's exploitation for traction, the steam engine dictated nationally a common form of economic activity and had created the conditions for the emergence of the factory system.

The final chapter of the story of steam relates to the development of the railway system. William Murdock had already built a model steam locomotive in 1781 which did eight miles an hour and at the same time experiments were being conducted in coalmining areas to improve the ways of moving coal away from the pit head. Pig iron rails were laid at Coalbrookdale for twenty miles and were also used to replace wooden rails in the Newcastle area. But the wagons were still drawn by horses. However, the important development took place of transferring the flange keeping the wheels in position from the rail to the wheel.

Although the Watt/Boulton partnership had the invention of the traction engine in their grasp it was another inventor, George Stephenson, who built the first commercial engine - not surprisingly for use in connection with transporting coal. But developments were slow, reflecting a lack of imagination. The Stockton and Darlington railway, sanctioned by an Act of Parliament in 1821 was half built before steam power was adopted. Even then the engine, as elsewhere, was used as a stationary engine to pull wagons up certain gradients. In 1830 the Liverpool/Manchester line was opened to transport both raw materials and fuel; this was followed by the Leicester/Simington line constructed to carry coal. In 1838 the Great Western line from London to Bath and Bristol was opened and this was followed by a massive expansion. There were 500 miles of track in 1838 and 5,000 miles a decade later.

The development of the steam engine and the subsequent construction of the railway network, represents the exploitation of what we might call a core technology; its history amply illustrates the set of stages described earlier that invariably accompany the introduction of new technologies.

First, despite a retrospective view of the revolutionary nature of the steam engine, it is amazing how slowly the invention was applied. Savery's invention took place in 1698. A century passed before, in 1781, the patent for a rotary motion was taken out. During that whole period the object of application of the new technology was to replicate the process of the existing water powered technologies. So we find the steam engine simply being used to raise water in order that it could be allowed to fall by gravity as with the action of the stream or river.

It was, however, clear that the new technology offered marginal advantages over existing water powered systems. There was greater efficiency, more power, wider applications (such as in the mines) and lower costs. Much of the task of exploitation thus went into improvements such as increasing the heat/output ratio of the machine, refining the manufacturing technology, etc., as well as adapting the technology for use in a growing number of different industries.

In the case of the steam engine, the fullest potential of the technology occurred when rotary motion was invented. Henceforth the technology ceased to be one related to pumping water, but opened up the wider vistas of potential for traction engines.

But even so, in order to exploit the advantages of the traction engine completely an infrastructure had to be created. In this case this was the task of laying track to create ultimately a national railway system. We have seen how at this stage, too, there was a significant degree of myopia and uncertainty as to the precise means of exploiting traction-based technology - a reluctance to abandon concepts of travel associated with existing systems. However, once the breakthrough had occurred, and the railway network had been built, the use of railways reduced the cost of transport of materials and products, opened up new markets and generated new industries such as the national newspaper market and tourism. Additionally, railway investment became a profitable use of surplus capital both at home and abroad.

The exploitation of the steam engine is a clear example of the enormity of change associated with a new core technology. By the mid-nineteenth century the impact of steam power, especially for traction, was pervasive; transforming a range of industries and revolutionizing working conditions. In association with other technologies, it created a period of vast economic and social change.

The century from 1750 to 1850 is a tantalizing one to the economist. The period covered the most significant upsurge in general economic activity that the world had hitherto known. Yet the nature and magnitude of many of the key changes, especially at the beginning of the period, are shrouded in statistical mists. The changes revolutionized the economy and the social framework itself - with fundamental repercussions - so we must persevere and make the most of what data we have got. We do so in the hope, not that we can magic out of the air comprehensive economic data, but that we can find key indicators that tell us their message and, hopefully, act as a safe proxy for those we cannot find.

To embark on this exercise we need to be sure of what we are hoping to find. Since we can safely accept the vastly different state of the economy in 1850 compared with 1750, we are looking for evidence of the way in which changes in technologies have transformed economic activity, and have cascaded through existing systems, destroying much in their wake, as well as opening up new vistas of opportunity; Schumpeter's process of 'creative destruction'. [3] As he describes it, 'that process ... that incessantly revolutionizes the economic structure from within, incessantly destroying the old one, incessantly creating a new'. In other words he regards it as a form of evolutionary process in which the fittest (new technologies) survive at the expense of the weakest (old technologies). Schumpeter actually uses the term 'industrial mutation'. Our interest, in that sense, lies therefore as much in the evidence of compositional changes as in the overall aggregate data - notwithstanding the importance of the aggregate message.

Schumpeter's useful classification has a special application to the first part of the period. In the initial stages new technologies were instrumental in the process of introducing an industrialized economy at the expense of an agrarian one. The process may well be of the same nature as later developments in the economy, but the scale of what was happening socially as well as economically puts the change into a somewhat different perspective. [4]

Deane and Cole [5] estimate that national income grew from £48 million in 1688 to £232 million in 1801, a rate of growth of approximately 1½ per cent per annum. But other indicators of economic activity clearly suggest an acceleration of industrial production from 1750 onwards. The consumption of raw cotton by British mills rose from 8,000 tonnes in 1760 to 25,000 tonnes in 1800; a rate of growth of just under three per cent per annum. Coal output was estimated to have grown by forty per cent between 1750 and 1790, but over the next five years alone output grew by thirty per cent. The output of pig iron grew from 25,000 tonnes to 68,000 tonnes between 1720 and 1788. By 1852 output stood at 2.7 million tonnes.

However sparse these data are, they do indicate a significant increase in economic activity, in output and probably in output per head (not necessarily reflected in the growth in real wages). The estimates of the latter are difficult to interpret, but they do suggest some improvement during the period although the high inflation rates of the Napoleonic war period make the turn of the century an uncomfortable reference point. [6]

From 1800 to 1850 the data are more reliable and more comprehensive. Hobsbawm [7] quotes the following figures for the growth in UK industrial production.

Table 2.1
Growth in UK industrial production

	%	% per annum
1800-1810	22.9	2.1
1810-1820	38.6	3.3
1820-1830	47.2	3.9
1830-1840	37.4	3.2
1840-1850	39.3	3.4

They show overall rates of growth comfortably over three per cent for most of the first half of the century; increases confirmed by other key industrial data. Use of raw cotton by British mills rose from 25,000 tonnes in 1800 to 300,000 tonnes in 1850. Iron output rose from 250,000 tonnes in 1800 to two million tonnes in 1850 and coal production increased from six million tonnes in 1770 to approximately fifty-five million tonnes in 1850. Evidence relating to real wages, however, suggests a decline of 0.5 per cent per annum in the decade ending in 1815 (part of the period of the Napoleonic wars) and a rise of 0.7 per cent per annum in the decade ending in 1847.

Deane and Cole [8] estimate changes in industrial composition in Great Britain (in terms of net output) as follows

Table 2.2
The composition of net output

	1688*	1801	1851
Agriculture, forestry and fishing	40.2	32.5	20.4
Mining, manufacturing and building	20.6	23.4	34.3
Trade and transport	11.7	17.5	18.7
Rent of dwellings	5.2	5.3	8.1
Other	22.3	21.3	18.5
Gross national income	100.0	100.0	100.0

* England and Wales only

This shows a massive swing away from agriculture where the contribution to national output halved over the century and a half. The bulk of the decline occurred between 1801 and 1851. The gainers were mining and manufacturing showing virtually a fifty per cent increase in share between

1801 and 1851, and trade and transport where the increase was largely in the earlier part of the period.

The same shift in pattern is revealed in employment data, as is shown by the following table (Table 2.3). [9]

Table 2.3
Numbers in employment by industry

	% 1811	% 1851
Agricultural occupations	35.2	21.9
Trade, manufactures	44.4	48.0
Other	20.4	30.1
Total	100.0	100.0

By 1851 the workforce within manufacturing was distributed as follows: [10]

Table 2.4
Distribution of workforce within manufacturing

	%
Metal manufacturing	15.8
Wood and furniture	4.4
Bricks, cement etc.	2.5
Chemicals etc.	1.3
Shoes and leather goods	1.7
Paper, printing etc.	2.2
Textiles	35.9
Clothing	25.2
Food, drink and tobacco	11.0
Total	100.0

It is significant that the bulk of this labour force was employed in consumer goods industries (and increasingly for export).

These data confirm the transformation that occurred in the British economy between 1750 and 1850. It had developed an economic potential which was able to sustain historically very high rates of growth in output, an acceleration that began at least by 1800 and probably before. Output and employment indicators both confirm the change in the distribution of net output largely as a result of the swing away from agriculture to manufacturing.

In terms of the location of industry two significant features stand out. The first is its concentration on urban areas. In 1751 London was the only city with a sizeable population (approximately 750,000). The next largest was Bristol (60,000) but no other city had above 50,000 inhabitants. Manchester had only 17,000 inhabitants in 1760. By 1831 Manchester had grown to 238,000, Leeds 123,000, Liverpool 202,000 and Glasgow 193,000. The growth was thus not around traditional cities such as Birmingham but in quite new areas.

Second, the introduction of steam power freed manufacturing industry from the constraint of the need to be close to sources of water power. However, most of the new industries continued to develop in the northern part of Britain because the location of coal and iron resources and the particular climatic conditions favoured textile production either side of the Pennines. The major sea ports of Liverpool, Bristol and Glasgow developed as a result of overseas trade; responding to the need to bring raw materials into a port as close to the manufacturing area as possible and also to export finished products to the growing empire.

By far the largest export from Great Britain in 1750 was that of woollen yarn and manufactures. Exports of woollen yarn made up roughly seventy five per cent of total exports. There was a small but significant trade in iron and steel and non-ferrous metals, and in related products that accounted for about sixteen per cent of total exports. The rest was made up of exports of coal and yarns of other textiles such as cotton, linen and silk.

By 1829 the pattern had changed radically. Exports in total had expanded almost tenfold (in money terms), but the dominant sector in terms of products was now cotton yarn and manufactures (making up seventy four per cent of the total). Exports of other yarns were relatively small. Wool took only eleven per cent and linen six per cent. Exports of iron and non-ferrous metals and related products, whilst growing substantially in absolute terms, declined relatively in the face of the phenomenal growth in cotton exports.

Table 2.5
Exports from Great Britain [11]

		%
	1750	1829
Coal	2.6	0.6
Iron and steel and manufactures	7.5	5.5
Non-ferrous metals and manufactures	8.5	2.6
Cotton yarn and manufactures	0.3	74.4
Woollen yarn and manufactures	75.2	10.7
Linen yarn and manufactures	3.9	5.7
Silk yarn and manufactures	2.0	0.4
Total	100.0	100.0

If the magnitude of the change brought about by the first hundred years of the industrial revolution is to be fully established, we need to supplement the evidence of underlying per capita growth in real income, together with increases of output in overseas trade, by evidence of rates of growth of the components of domestic demand. Unfortunately comprehensive data about domestic demand for this period are not available.

Public expenditure changed in shape and size over the century beginning in 1750. Starting from a pre-industrial pattern where a large proportion of public expenditure took place at local level (with defence expenditure being the largest central government item), by 1850 the inheritance of the Napoleonic wars was a much more complicated public financing pattern. Central government expenditure was committed more heavily to defence matters and now had the task of servicing the debts run up to finance the war. Government was by now, also, becoming more involved in other civil expenditure associated with servicing social programmes - in education, health, safety supervision, the administration of the Poor Law etc. By 1851 the state provision of services was already becoming a key element in the overall pattern of demand.

The rapid acceleration in the application of new technologies already described, required significant capital expenditure. The difficulties of the Watt/Boulton partnership in launching the steam engine have already been noted. The expansion of the railway system from 500 miles in 1838 to 5,000 in 1848 is also an example of a development requiring heavy investment. The precise overall levels of capital formation during this period are not known. However, many of the changes introduced up to, say, 1825 probably did not demand excessively heavy investment; that was to come later. The earlier technological advances (particularly in textiles) were basically simple and relatively cheap to implement - their cost was easily accommodated by the surplus funds that had already been created by the earlier period of commercial capitalism. Indeed, such was the success of the first phase of the industrial revolution that it created the levels of investible funds required for the much more complex and financially demanding later stages of development of the middle of the nineteenth century.

The most elusive changes in the economic map of the century from 1750 that would tell us so much about the impact of the revolution, if we could lay our hands on appropriate data, would be an indication of changes in the size and composition of consumers' expenditure. Unfortunately, this area is equally difficult to examine. The starting point is relatively clear. In 1750 patterns of expenditure for most people reflected the nature of the economy they lived in. Their's was a world of small towns and a local agricultural catchment area surrounding the town, with hardly any trading between urban areas. Consumption for the masses still consisted of getting access to basic commodities such as food and drink, clothing and footwear; their production would be essentially local. In some cases barter may still have prevailed. Most trading was face to face and the commodities traded were produced in adjacent or facing shops, traded between neighbours and between town and farm.

There were important exceptions to this pattern that add to our information and give us clues why Britain was particularly fertile ground for the launch of the industrial revolution. Hobsbawm [12] points out that by 1750

> ... 0.6 pound of tea was legally imported per head of the population, plus a considerable amount smuggled in, and already there was evidence that the drink was not uncommon in the countryside, even among labourers (or more precisely, their wives and daughters). The British, thought Wendeborn, consumed three times as much tea as the rest of Europe put together.

These statistics endorse the view that there already existed a market economy in respect of certain goods, a development aided and abetted by the use of cash - both as payment for labour and as a means of exchange. It would be wrong to conclude that in 1750 there was a mass consumer market anywhere of the kind that developed later except in the relatively close circles of the privileged rich and middle classes.

Of more interest would be evidence of the consumer market of 1850. Such evidence would tell us much about mid-Victorian attitudes towards consumerism and, by inference, to their understanding of what it meant to be involved in the market capitalist system and how that affected their attitudes to work, leisure and consumption. We tend to think of the mid-century Victorian as someone committed to John Wesley's maxim of 'earn all you can, save all you can and give all you can'. This highly virtuous view of economic activity gives no sign of the later view which we have inherited and is probably better summed up as 'earn all you can and spend all you can'. But was this virtuous view of the mid-Victorian true? Unfortunately, we have no measures of consumers' expenditure patterns on a comprehensive basis so in the main evidence has to be anecdotal. One event can tell us a great deal about the changes between 1750 and 1850.

The Great Exhibition of 1851, with its 15,000 exhibits, was an emphatic testimony to the progress that British industry had made over the hundred years since the beginning of the industrial revolution. The fillip that the exhibition itself gave to the mid-century economy was considerable and included the consolidation of railways as a mass transport system, particularly around London, and the stimulation of mass consumerism (William Whiteley went on to found the first department store a decade later). The exhibits themselves demonstrated the importance of the consumer market and the role of British heavy industry in supplying the rest of the world with capital goods. There was a sense of awe experienced by the Victorian visitor as recorded by Christopher Hobhouse.

> We enter the building to see the largest sheet of plate glass ever made and a miscellany of mineral ores and foodstuffs; also a model of the Liverpool docks (Britain the great metropolitan consumer). Machinery, headed by locomotives and other railway equipment: James Watt of Soho: the hydraulic press which lifted the Britannia tubular bridge: Mr Shillibeer's expanding hearse. Manufacturing machines and tools - James Nasmyth's steam hammer easily first, being so big and so gentle. Bridges and lighthouses. Ship models and life belts. A prodigious assemblage of sporting arms... agricultural implements, drawing a rural crowd, which to Mr Hobhouse is more attractive than the implements themselves. Among philosophic instruments the great Ross telescope and some photographic apparatus; balloons of course, for it was the age of balloon flights, and musical instruments galore. Pianos by Broadwood, Collard, etc. Clocks and watches and surgical instruments which strove to outdo one another in fantasy. [13]

Section followed section of exhibits. First those devoted to textiles, precious metals and jewelry, pottery and furniture and sculpture. Then those containing machinery, followed by others covering minerals and other commodities, as well as all sides of manufactured products designed for the emerging consumer market.

There were, of course, sections devoted to foreign and colonial products. There was no embarrassment felt in their presence. Britain was quite content to see them alongside British products. 'It was because Britain believed she had things to offer them, the rest of the world, which, whether the world wanted them or not, were destined to enter into its consumption, that she bestrode with such nonchalance the free trade horse'. [14]

If we once started listing exhibits in full detail we would not know where to stop, but a brief summary will underline the range and quantity of the output of the formidable industrial machine that had been created in Britain by this date. The list of machines included, for example,

a. Steam engines and boilers, water and windmills and various other prime movers

b. Separate parts of machinery, specimens of workmanship

c. Pneumatic machines

d. Hydraulic machines, cranes, etc.

e. Locomotives, railway carriages

f. Railway machinery and permanent way

g. Weighing, measuring and registering machines

Other machinery included machines for the manufacture of metal goods, minerals and other substances. Civil engineering covered foundations, scaffolding, bridges and tunnels, docks, roofs, water, gas and sewage works and heating equipment. The section for military building (including ship building) was understandably impressive.

In the field of consumer goods, there were large sections for cotton, woollen and silk fabrics and made-up goods, many items of which, such as 'ginghams, hungarians' leave much to the imagination. There was a large section for leather and leather goods, paper and paper goods, fabrics presented as specimens of printing or dyeing, tapestries and other articles of clothing. These were followed by sections containing cutlery, iron and general hardware and products made from precious metals, glass and ceramics, as well as furniture and other miscellaneous products.

The exhibition represented a stocktaking of what had been achieved in the first century of the industrial revolution. It was a formidable performance which Britain was proud to show off to the rest of the world. But there was also an inner pride in the emergence of what we would now call a mass consumer market. In many respects the exhibition marked a beginning rather than an end, for the pace of expansion of the consumer market began to quicken. The range of goods continued to expand, but the exhibition also provided the stimulus to new forms of retailing for the mass markets. As mentioned already, William Whiteley mirrored his department store layout, in Paddington, London, on the open displays at the exhibition and subsequently trained the men who went on to found other major London stores, such as Harrods, John Barkers and the Army and Navy Stores. The exhibition also enabled the railways to service the mass travel market - helped (who else?) by a young Thomas Cook who introduced the railway excursion ticket (including the price of admission to the exhibition) for the Midland Railway.

Prince Albert prayed 'that the exhibition of 1851 may prove in its results to have been the means of advancing the happiness and prosperity of not only this (Britain) but of all other countries'.

In retrospect this was not an exaggerated claim. The exhibition made Britons totally aware of the impact of the new industrial age where British industrial supremacy was taken for granted. It was a clear measure of the distance Britain had travelled over the hundred years since 1750.

It would be idle to imagine that economic changes of the magnitude already described would have had little or no effect on the wider social structure. On the contrary, the changes struck at the fundamentals of society itself; its infrastructure and the values upon which it was based. In contrast to the task of documenting economic change over the century from 1750 to 1850 where so little systematic information is available, social changes, and particularly the adverse effects, have long been the object of enquiry for historians and there is much to draw on. In contrast, whilst the broad beneficial effects are there to be seen in the economic data, it is less easy to get a picture of the everyday life of those who gained from the process. Those who willingly embraced new styles of living, and indeed new values, upon which to base and justify their participation in the new venture of economic growth.

One of the major dislocating factors created by the acceleration of technological and economic change was the movement of population. The great movement off the land was illustrated in Table 2.3 which showed that the proportion working in agriculture falling from 35.2 per cent in 1811 to 21.9 per cent in 1851. But the dislocation was greater than this. What emerged was a movement into factories which, because of the long hours demanded of the labour force, required a proximity to work that inevitably meant a concentration into towns and cities on an unprecedented scale. Towns with a population of 50,000 or over became commonplace; they generated community challenges with which the existing local government structure was incapable of dealing. The local government structure depending largely on the parish system quickly collapsed in those areas where the industrial population grew rapidly.

The movement of labour provoked by industrial change led to other pressures being generated, which also stretched existing structures to breaking point. The most important of these was the increase in the movement of the population from one region to another over far greater distances than before to look for work, particularly at times of recession. The widening impact of change owes much to the rapid improvements in the means of transportation. By 1850 it was possible to travel by rail from Manchester to London in seven hours and from Edinburgh to London in twelve hours. Thus the labour force took on a mobility with which the local system of government, essentially restricted by the parochial system, was quite unable to cope.

Although the precise cause and effect continues to be debated, the population of Britain also began to increase rapidly after the turn of the century. The total population in 1750 was 6.5 million, by 1801 it had risen to nine million and by 1841 it was sixteen million. Whether a Ricardian response to higher living standards or not, the larger numbers added to the sense of radical transformation throughout the land.

Much has been written about the lives and lifestyles of some of those who demonstrably gained from the industrial revolution. Many made fortunes and their rewards were reflected in large houses, large retinues and, it seems, in a great deal of public ostentation. Many of those who gained had already been well established prior to 1750. Many landowners readily saw the commercial opportunities, for example, of mining under their soil or of selling land for industrial or railway development. The aristocracy, by providing much of the funds for the early stages of industrial capitalism, clearly gained and, of course, so did those engaged in activities feeding off what would now be called the establishment; the universities, the civil service, the church and the professions - not the least being the expansion of the public school system to meet the huge increase in demand for public school education; the route by which men of commerce and trade sought to gain respectability for their children and their families.

The trends that are the least easy to identify are those relating to the impact on the newly created industrious, entrepreneurial, middle classes. It is certain that this group provided the dynamic for the industrial revolution, were the main recipients of its benefits and had the greatest influence on Victorian society. Their impact was economic as reflected in the pursuit of laissez-faire economic policies and in the creation and control of the factory system. Their impact was also political by virtue of their support for the great reforming Acts of Parliament that emerged as the century unfolded. However their influence was also on lifestyles and social behaviour, since the dynamics of the industrial revolution could not have been sustained without the pressures exerted by the new found mass consumerism. Their influence was also ethical, to the extent to which this group embraced and practised what might loosely be called the protestant work ethic.

To a substantial extent the new managerial/entrepreneurial classes sought to ape their 'betters'. The houses were smaller, the retinues were smaller, but the lifestyle was much the same. The aspirations for social status and to secure the future for their children were the same. Weiner [15] makes an important point in arguing that the new classes never fully embraced an industrial culture in the sense that Germany and the United States did later, since the lifestyle and sets of social distinctions sought after remained those valued by pre-industrial society.

In a sense, the lot of the class of people just below the new entrepreneurs and managers provides more food for thought. The rapid development of the industrial revolution presupposes a vast expansion in the numbers of workers, skilled, semi-skilled or unskilled and in junior levels of management. Although there were bad as well as good times we can assume an underlying material improvement throughout the century in the standards of living of these groups. It was these groups, therefore, that provided the main domestic stimulation to economic activity and which provided the source and object of consumer innovation as the century went on. The gradual increase and extent to which the economy became dominated by consumer expenditure and therefore, in a sense, populist, gave the industrial revolution the final stamp of critical difference to the past.

Of course, consumerism as we know it now was a long way off. But it suffices to note that the power of the masses, which was reflected in consumerism, was also expressed in trade union and political movements; the latter being equally decisive in determining the share of the emerging modern economic structure. It is possible to argue that the developing dependence of the economy on consumer demand created a unifying force, an identity of interests, between management and labour which served to mollify otherwise divisive political trends.

There were thus many gainers from the radical transformation of the economy. These gains have to be measured in monetary terms, but also in the sense in which each successive generation saw a lifestyle opening up in front of it to which it readily aspired and which it greatly enjoyed.

The emphasis on those who gained is not intended to detract from the weight of evidence that can be marshalled to demonstrate who lost out in the process. If we assume that the historians have sifted the evidence correctly then the cost of industrialization in terms of suffering, both economic and social, was substantial.

Many of the criticisms concern the evils of applying laissez-faire economics - evils in the sense that the imposed working practices and conditions were seen to be an affront to conscience. The fact that many of the evils were a long time being overthrown tells us much about the speed at which perceptions of justice change and develop in a society.

However, with the comfortable benefit of hindsight, the conditions under which, for example, children were obliged to work in the 1830s were horrific. For children [16] to have to leave home at 3.00 a.m. and not return until 10.00 p.m. each day for at least six days a week, with only fifteen minutes each for breakfast and drinks and thirty minutes for dinner (with duties to perform even during these breaks), would now be regarded as quite intolerable. Similar stories can be told of evidence that led to the struggles for increased safety at work, for the protection of women at work, for the reduction of working hours and the fight for minimum wages.

The intensity and duration of these reforming campaigns underline the pain and cost that was initially incurred by the process of change. But there was no greater example of both the price being paid and the inability of the existing structures to cope with it than society's attitude to the plight of the poor. Even for those in employment the conditions in the cities must have been pretty bad. 'Civilization works its miracles and civilized man is turned back almost to a savage', wrote the French writer, de Tocqueville, of Manchester in the middle of the century [17]. The poverty in the cities was accompanied, too, through the emergence of separate living areas, by the creation of a social separation between rich and poor, the haves and have nots, which was never present in the old agricultural society.

It was the treatment of those without work that occupied the time and thoughts of so many. The poverty of the industrial revolution was new. There continued to be those who could not work - the ill and the handicapped - and also those who would not work - the lazy and unmotivated. These were the poor that the parish system had previously dealt with more or less adequately. Each parish owned its responsibility to its parishioners. Poor relief out of parish funds was available to the deserving. The lazy were chastized. What threw the traditional system into disarray were two factors that were eminently consequences of the new economic regime. The first was the new degree of enforced labour mobility so that parishes were no longer faced with the lesser problem of dealing with the poor of their own parish. They were now being swamped by strangers coming into their parishes either in work, to seek work, or just out of work. The numbers involved made it impracticable to send those without work back to their original parishes. To do so, in any case, would have attempted to move against the currents of the time which demanded labour mobility. Employers or justices of the peace could hardly complain that destitute strangers were turning up in their parishes when they were glad to have those strangers that they were content or able to employ.

Second, the old system could not cope with the other new phenomenon of economic growth. The new economic system, in a way far greater than any previous system, turned out to be impossible to control in a way that provided steady employment (if not full employment). Not that such a policy objective was even considered at that time as one for the government of the day to work towards, save in the sense of trying to alleviate conditions in particular industries. As Rostow [18] decisively demonstrated, the economy created by the industrial revolution was subject to cyclical swings featuring a business cycle of five years duration (perhaps also accompanied by waves of longer duration) which has continued to be a feature of the type of economy born of the heat of the first technological surge.

The result was a new form of unemployment; those who wanted to work but could not find it. For a considerable while, indeed for most of the nineteenth century, the Poor Law system clung to the belief that the responsibility for this matter rested with the individual worker. Had he searched for work hard enough? Or, more importantly, had he offered his labour at the appropriate price for the market to take up his offer? This latter view - that there was a price which would clear the market (for anything, including labour) - was held solidly throughout the nineteenth century. It firmly placed the responsibility on the individual concerned to take whatever initiative was required to restore the market equilibrium (as a result it led to the doctrine of less eligibility, which set the level of Poor Law support always less than the lowest prevailing market wage rate). Even in other circumstances where individuals were clearly not to blame for their condition, such as the potato famine in Ireland in the 1840s, the same rigid adherence to non-intervention was applied.

The task of coming to terms with the burden and nature of poverty became one of the most significant reforming movements of the nineteenth century and into the early twentieth century. It was important - perhaps ranking above all else - for it struck at the heart of the culture of the new society. Its existence as an intractable problem served continuously and ominously to qualify the heady sense of success, of triumphant progress that the Victorians wanted to celebrate. As a pertinent and shrewd contemporary observer of the mid-nineteenth century described the dilemma,

> This association of poverty with progress is the great enigma of our time. It is the central fact from which spring industrial, social and political difficulties that perplex the world and with which statesmanship and philanthropy and education grapple in vain. From it come the clouds that overhang the future of the most progressive and self-reliant nations. It is the riddle that the Sphinx of Fate puts to our civilization, which not to answer is to be destroyed. So long as all the increased wealth which modern progress brings goes to build up great fortunes, to increase luxury and make sharper the contrast between the House of Have and the House of Want, progress is not real and cannot be permanent. [19]

From the end of the first quarter of the nineteenth century a process of reform began which gradually permeated the whole of British society. The process was often based on what has been called 'Blue Book sociology', a title given because many of the major reforms requiring Acts of Parliament were initiated in roughly the same way. The awareness of the need for reform began with the mounting social and political pressure growing to a point where parliament was unable to ignore it. The immediate response was the creation of a Royal Commission charged with the task of collecting evidence to confirm or reject the allegations that were being made. The Royal Commission report contained the evidence collected which, in most cases, broke new ground in terms of measuring and assessing the social conditions that were being addressed. The publication of the data in the Royal Commission's report (always published within blue covers - hence the term 'Blue Book sociology' - a practice that continues to this day) invariably struck at the nation's conscience and led, sometimes quickly, sometimes slowly, to reforming legislation.

An example of one such report, already referred to, was that of the Select Committee on Factory Children's Labour published in 1832 based on evidence taken during 1831 and 1832. In retrospect the evidence of cruelty and inhumane conditions appears overwhelming, but, besides this evidence of suffering, the report also contained the arguments for and against the use of child labour. It was by no means a foregone conclusion that changes ought to be made. The two principal arguments deployed against change were firstly that a reduction in hours would increase wage costs and make products uncompetitive, a powerful argument in the period when pure laissez-faire doctrine was sweeping through the economy. The second argument was more subtle. It was that to curb the hours that children could work would be an infringement of the child's basic right and freedom to work as it wished. However, the combined pressures of the reformers, humane manufacturers and religious groups, together with landowners glad to support an attack on the industrial sector, as well as adult workers themselves, won the day. The consequent reforming legislation was slow in coming and bitterly contested, but eventually a series of Factory Acts, of which that in 1833 was crucial, introduced regulations which eventually embraced working hours for children (and subsequently for adults after 1850) and health and safety at work. However, the widespread application of such principles had to wait until the second half of the century before it reached most industries.

Two other areas that embodied a reaction to the conditions created by the first phase of the industrial revolution, were poverty and public health. The contemporary public view of poverty prevailing through the period 1750-1840 (and beyond) is well known and well documented and has already been referred to and will be addressed again later. Essentially the view that prevailed for so long was that the undeserving poor, i.e. able bodied men without work, should be forced to find work by being presented with the far more unsavoury prospects of living under workhouse conditions, separated from family and by being paid money which was less than the lowest prevailing rate outside in the real economy.

The view that the responsibility for finding work, or getting back into work, rested entirely with the out of work man was rooted in deep ideological views. The alternative view, that the unemployed were often the victims of technological or cyclical forces beyond their control, barely attracted any credibility until well into the twentieth century. The factors that eventually undermined the Poor Law system, as it was known, were firstly the realization that however culpable the adult male was perceived to be in failing to find work, his wife and children could clearly not be blamed to the same extent. Eventually this view made it intolerable that families should be taken out of the community and into institutions. Slowly, therefore, giving relief outside the workhouse was increased but children continued to be punished for the 'sins' of their fathers, often living in mixed workhouses and attending workhouse schools (even once school attendance had become compulsory in 1870).

The factor that did most, however, to undermine of the Poor Law system was the problem of the treatment of the sick, physical and mental, and the elderly. The workhouse remained the focus for treating these categories but gradually it became apparent that medical attention of one sort or another needed to be provided. The growth of hospital wings attached to workhouses, which ensued eventually, provided a key building block in the subsequent creation of a hospital system in the twentieth century. The desire to keep the elderly in workhouses was less strong. Pressures soon mounted for all elderly couples to stay together and to have more congenial accommodation. Eventually the Old Age Pension Act of 1908 provided pensions which were not subject to the stigma of poor relief.

The administration of the Poor Law Act was thus already raising public health issues, but these were more widely assessed by a number of other Blue Books, notably the Report on the Health of Towns Committee published in 1840 and another sponsored by the Poor Law Commission and written by Edwin Chadwick was published in 1842 as the report on the Enquiry into the Sanitary Condition of the Labouring Population of Great Britain. These revealed the public health threat of city overcrowding, insanitary disposal of waste, tainted water supplies and the ease with which infectious and contagious diseases could spread. As disease is no respecter of person or class, it is understandable that the middle classes, and their friends in parliament, moved relatively quickly to improve the conditions revealed by these enquiries.

The reforms introduced to deal with these issues led directly or indirectly to the creation of the major state institutions that are now commonplace, reflected in the local government network and its supervision from the centre, and the major departments of health and social security.

However, much of what was eventually done to improve public facilities derived from and also enhanced the great sense of civic pride that developed from the middle of the nineteenth century on. The practical applications of this sense of pride are still to be seen in the splendid Victorian civic buildings throughout the major cities of Britain; this was only part of the new social infrastructure created in response to the pressures and stimuli emanating from the industrial revolution. Virtually all existing institutions were reformed to make them more appropriate to the needs of the time, including the legal system (the Law Society was first founded in 1831), law and order, education and health (the British Medical Association was first founded in 1854). Other facilities, professions and skills emerged for the first time. The growth in management skills and its subsequent documentation to make it an academic area of study, is one notable example. So, too, was the flourishing of learned societies of which the Royal Institution (created in 1799) was outstanding. It was established 'for the diffusing of knowledge and facilitating the general and speedy introduction of new and useful mechanical inventions and improvements, and also for teaching, by regular courses of philosophical lectures and experiments, the application of these discoveries in science to improvements of arts and manufactures, and facilitating the means of procuring the comforts and conveniences of life'. The Royal College of Chemistry (and with it the Institute at Rothampsted) was founded in 1843 and the British Association for the Advancement of Science in 1831.

We must not forget, too, the many pleasures and benefits that came to those who gained from and were comfortable in the new order. We have already noted the increased access to consumer goods - a far cry from the early Ricardian view that workers worked best when struggling for subsistence. But many other social trends converged to produce a strong sense of achievement and belief. It was to be found in a religion that underpinned the values upon which economic activity was based and in regular church going which provided an important outlet for socializing and for revealing self-esteem and self-worth. It was to be found in a strong sense of nationalism, owing much to the perceived triumphs of empire. It was also to be found in the degree of satisfaction, if not arrogance, that derived from a nation that felt it was first in the field of almost everything.

These very conditions, and especially the contrasting pain and hardship of the whole period, of course, produced its own dialectic which was reflected overtly in the political process both in parliament and in the political activities of the trade unions. It was also to be found in the many working class movements, often with utopian overtones, which are still there to be seen in the slogans found in the banners of these organizations. But it was also there in the underlying changes in what E.P. Thompson [20] calls custom - or in working class culture, to give it an alternative name. This dialectically created other world, was as clearly brought into being as a consequence of the process of technological change, and is as important an element in the new society, as the beliefs and practices of those who gained most.

However debatable the timing and however contentious the attribution of causes, the period 1750 to 1850 was one of rapid change. Of course, it did not happen all at once. In retrospect it is surprising how slow many of the inevitable consequences of the initial changes were in coming through; they did so eventually, with overpowering force. In many cases the response was much like a head of steam developing until eventually it reached the point of venting after which nothing looked the same again.

Undoubtedly the major pressures for change were to be found in the new technologies. Granted they emerged at a time when other conditions in Britain were favourable. The availability of surplus funds created by decades of successful commercial activity and successful war making, an inventiveness arising out of the flourishing scientific spirit of the age and a social structure particularly open to being shaped in new ways to accommodate and exploit the new changes, all helped.

The impact of the new technologies, particularly the introduction of the steam engine, was devastating. In its stationary form the steam engine was the core technology which allowed all sorts of other technology to be introduced; power driven looms, better mining techniques, iron and steel making. In its ultimate use for locomotion it quite decisively transformed the industrial scene.

The initial consequences were to destroy the previous social economy based on agriculture and domestic, farmed out, production. It was replaced by factories and all that meant in terms of the location of industry, new working conditions, the social consequences of rapidly growing industrial cities and the underlying contractual motivation for buying and selling labour. The system created an economy which in one sense was clearly an instrument for economic change and wealth creation, whilst in another sense it emerged as a taskmaster which, subjected to periodic trade cycles, could no longer guarantee the permanence of work which, for the most part, had been a characteristic of earlier systems.

The forces propelling the principal players in this drama - scientists and engineers, entrepreneurs and many privileged people who stood to gain - were powerful. A new spirit was in the air where success after success continually reinforced the air of confidence. Money making was good, winning the competitive struggle was fulfilling. The support for these forces came mainly from the new scientific environment developing in an atmosphere of confident humanism. The process of scientific enquiry, backed by enterprising inventors and technologists, was, by the end of the eighteenth century, ensuring a steady flow of inventions of immediate practical application. It was given more drive and more purpose in an odd way by being perceived as a battle with an old order of thinking represented by the church. The growth of secular thinking which, later, cut its teeth, for example in the great debate over Darwin's theory of evolution - where toppling the church became as important as defending Darwin - was a major force opening up Victorian society.

The growth in empire also contributed greatly to the ways in which the new forces broke through. Apart from the economic aspects of empire in guaranteeing markets, the empire opened up a two-way flow of both goods and culture that greatly reinforced the openness of the Victorian society to new ideas. To that has to be added the immeasurable boost to national confidence created by the success of empire.

Finally, the Victorians were able to lay their hands on a set of values, predominantly derived from evangelical Christianity, which provided them with a firm justification for acting and working as they did. This set of values, together with the intellectual power of utilitarian theories of Jeremy Bentham and others, provided the moral and ethical basis for the forces opening up the economy and society.

Despite the overwhelming strength of these new forces, both direct and indirect, society was not lacking in points of resistance. Many of these were bastions of the old order, others were being contemporaneously created in dialectic terms by the new forces themselves.

The experience of the defence put up by the elements of the old order in the face of the new forces makes a sad story. In historical terms the speed, for example, at which the new forces swept away previous working practices was breath-taking. Little of the social structure had a chance of surviving; village life, the village and small town economies, the rhythm of work itself, the values systems, the practice of right and wrong in personal and contractual relationships, the role of the state at both national and local levels all suffered. None of these key elements in the previous order held out in the end, and, given that the new systems were slower to come in place, a measure of the degree of suffering and uncertainty of the time is reflected in the nature and state of this interregnum. The feeling is probably to be compared to that experienced in Eastern Europe in the years immediately following the collapse of the command economy and its associated political system in our present day.

In a sense the growth of countervailing forces makes more interesting and encouraging reading. The stimulus for reaction, if one was needed, could be found there in the cultural and social interregnum, but evidence of movements designed to limit the excesses of the new forces very quickly appeared. It is wrong to see them as attempts to hark back to earlier times. On the contrary, most of these movements were very much the creation of the new forces and rapidly became as much part of the new society as those forces still wishing to push forward in an untrammelled way.

High on the list of these countervailing forces is the practical exercise of conscience. The history of the period from 1750 is rich with examples of great reforming personalities and movements, such as those associated with the names of Wilberforce, Chadwick, Owen and Fry to name a few. We must not overlook this continuous questioning of the moral consequences of events even if the nature of the diagnosis and the consequential proposals for reform were both heavily contextualized by the perceptions and values of the time. One of the main ways in which conscience was, so to speak, made manifest was in the stream of Blue Book sociology referred to earlier. Each of the major social issues, such as child labour, and the way in which each was dealt with, reflects the conscience of the age - a setting of boundaries outside which society was not prepared to allow the new forces to go.

Secondly, we should note the reform movements leading step by step towards our present form of democracy and parliamentary institutions. It is an interesting issue, which falls largely outside the scope of this study, why the reform movement, which eventually created the possibility (and fact) of mass political parties (including those of the left), developed the way it did. Why was there not an British equivalent of the French revolution? Perhaps it was because enough people asked that question at the time and were frightened by it, that an alternative political process, less revolutionary and more reforming, was set in train. However, the fact remains that reforms were introduced which were essentially against the interests of the class to which the reformers belonged and led (a point which is of particular relevance to us) to the creation of institutional elements which served to constrain the forces for economic and social change unleashed by the technological revolution.

The trade union movement is perhaps the classic case of the emergence of countervailing force which served to contain the new forces. Equally, a new interpretation and understanding of the values basis of society, provoked no doubt by the evidence of poverty and suffering, was developed within and outside the churches. Each of the major denominations (Anglican, Roman Catholic, Methodist) contained its own emerging social ministry (and movements to put it in practice) which, together with the reforming zeal of non-Christians, such as the Webbs, began to build up a reservoir of community based values. These provided the basis for community action and were put alongside the individualistically based values underpinning laissez-faire activity.

The education scene was more ambiguous. The system put in place was designed to support the new society and the new forces, either by inculcating values and the teaching of skills appropriate to the industrial life (or by helping to keep children off the streets whilst their parents worked if a more cynical view is permitted). By contrast, the great expansion in public schools reflected the desire on the part of industrialists to give their children an education more in tune with earlier perceptions of the role and purpose of education, which is now thought by many to have severely detracted from the flow of industrial leadership across the generations.

The process from 1750 to 1850 can thus be viewed as a period in which strong new pressures for change, technological in origin, were determined to break through. In the end they were unstoppable and once they were in full flood they swept virtually all before them. To start with, the new forces had to prove their worth, by fighting for a place within the context of an old framework. Having done so, they had only to wait for the creation of a new infrastructure, both technical, economic, social and more, to be able to exploit their potential to the full. Yet in this very process they created new forces devoted to economic and social containment by which the system became ethically justifiable again.

Notes

1. Toffler, Alvin (1981), *The Third Wave*, Pan Books, London
2. Ibid.
3. Schumpeter, J.A., op. cit. p. 83
4. Rostow, W.W. (1948), *British Economy of the Nineteenth Century,* Clarendon Press, Oxford, remains the outstanding example of an attempt to provide a comprehensive economic analysis of the nineteenth century.
5. Deane, Phyllis and Cole, W.A. (1962), *British Economic Growth 1688-1955*, Cambridge University Press, Cambridge. Mitchell, B.R. (with Deane, P.) (1962), quoted in *Abstract of British Historical Statistics*, Cambridge University Press, Cambridge, p. 366
6. Watson, J.S. (1985), *The Reign of George III, 1760-1815*, Clarendon Press, Oxford, p. 520
7. Hobsbawm, E.J. (1969), *Industry and Empire*, Penguin Books, Harmondsworth, p. 68
8. Mitchell, B.R. (with Deane, P.), op. cit. p. 366
9. Mitchell, B.R. (with Deane, P.), ibid. p. 60
10. Mitchell, B.R. (with Deane, P.), ibid. p. 60
11. Mitchell, B.R. (with Deane, P.), ibid. p. 294/5
12. Hobsbawm, E.J., op. cit. p. 28
13. Fay, C.R. (1951), paraphrased from '1851 and the Crystal Palace', *Palace of Industry, 1851, A study of the Great Exhibition and its fruits*, Cambridge University Press, Cambridge, p. 80
14. Ibid., p. 84
15. Weiner, M.J. (1981), *English Culture and the Decline of the Industrial Spirit 1850-1980*, Cambridge University Press, Cambridge
16. See evidence given to the Committee on Factory Children's Labour of 1832, extracts of which are given in Gregg, P. (1965), *A Social and Economic History of Britain 1760-1965*, George Harrap & Co, London, p. 121
17. Quoted in Hobsbawm op. cit. p. 86
18. Rostow, W.W., op. cit.

19. George, Henry (1879), *Progress and Poverty*, 1970 Centenary edition published by Biddles Ltd, Guildford, p. 5
20. Thompson, E.P. (1991), *Customs in Common*, Penguin Books, Harmondsworth

3 Technology and change in contemporary society

This chapter develops and consolidates the argument of the previous chapter. It comes forward in time to examine developments since 1960, since when it can be said that society has been locked into what can be called the new service economy. This choice of period offers two advantages. It makes it possible to examine, in another example, the nature of the revolution which results from the introduction of new technologies. It also provides, by examining the interaction between the economy and present day society, a backcloth to the immensely challenging contemporary issues facing Britain; both as a nation, as a member of the European Union, and as a member of the industrial west. That challenge includes framing the appropriate response to the formidable competition from the Pacific rim countries, to the transformation of those countries previously part of the hegemony of the USSR, and to the 'sleeping giants' of China and India, as they embark on the path of economic development.

Chapter Two concentrated, in terms of economic development, on two major factors, the rate of economic growth and the nature of the diffusion of change at sectoral level. By way of recapitulation, a useful way of approaching the issue of economic growth, and summarizing the nature of the economic problem is to introduce Eric Roll's [1] definition of the latter.

> It follows that the basic concept of the surplus product or value simply means that human labour is able to wrest from nature more than the bare needs for human survival; that all progress (and civilization itself) depends upon the size of this surplus; and that the division of this surplus between consumption and accumulation and among the various members (or 'classes') of the community is a central economic problem determining, to a major extent, the development of the economy itself.

Formulated in this way the definition introduces the key issues in the relationship between technology and economic growth, and the factors that determine a society's ability to increase its surplus. It points up the vital issue of the trade-off between consumption and investment, and addresses the most thorny issue of all, the question of income distribution over which various ideologies have taken their stand, especially as to 'who owns the surplus?'.

As regards economic growth the more humble amongst academic economists would readily admit that little is known about its causes. It is not through lack of effort since the question figures very large in economic theory. That theory is adequately represented in the concept of the production function which seeks to explain changes in output in terms of changes in the quantity or quality of factor inputs, such as labour, capital and land.

The construct is fundamental in theoretical terms but enormous problems of definition and measurement are encountered in the attempt to quantify the contribution of the respective causal factors. It is difficult to measure the factor inputs especially quality changes in them, such as the effect of better trained employees, or of newer equipment. Nor does the relationship adequately embrace the contribution of factors such as personal management skills, management systems or the role of industrial and labour organization. The difficulties created by all these factors are reflected in the fact that in most attempts to fit a production function the residual item (usually expressed as a time variable and taken as a proxy for technical progress in its broadest sense) often 'explains' more than half of the recorded changes in output.

Leaving aside the complexities of the analysis of production functions, it is worth noting that average overall rates of output growth have stayed within remarkably narrow bands over a long period of time, see Table 3.1.

Table 3.1 [2]
Rates of growth in output

Years	Rate of Growth % per annum
1856-1873	1.3
1873-1913	0.9
1924-1937	1.0
1951-1973	2.4
Average 1856-1973	1.2

The differences have not been great, although it must be remembered that a rate of growth of 1%-2% over, say, one hundred years represents a huge increase in material prosperity. But this apparent stability in the large aggregates, hides the key element in technological change, which is the speed and direction of the diffusion of new technologies. Over time this process has led to massive changes in industrial composition. Chapter Two illustrated the fact in respect of the changes between 1750 and 1850. The following table illustrates how the process is continuing. Service employment, just over fifty per cent in 1971 now accounts for almost three-quarters of all employment.

Table 3.2
Employees in employment - by industry (%)

	1971 (mid year)	1993 (mid year)
Agriculture, forestry and fishing	1.9	1.2
Mining and energy	3.6	1.7
Construction	5.4	3.9
Manufacturing	36.5	20.4
Transport and communication	7.1	5.9
Wholesaling and retailing	16.7	21.5
Financial services	6.1	12.7
Education and health	10.2	16.1
Public administration	8.0	8.5
Other services	4.5	8.2
Total services	52.6	72.9
Total employment	100.0	100.0

In going on to examine the impact of technological change today it is worth repeating the approach adopted in Chapter Two which distinguished four separate stages in the introduction of a new technology. The initial stage occurs where a new technology is first introduced in the market to compete with well established older rival technologies; to do so it has to reproduce as best it can the qualities and familiar characteristics of the products it is seeking to replace. At this stage the similarity of the products has to be emphasised in order to gain access to a well established market. In the second stage the new technology has to demonstrate that, besides being able to do all the things that products using the old technology were able to do, the new technology can provide valuable and desirable but still marginal extra attributes. The third stage is where, having established itself in the market, it is quite clear that to gain the full benefits of the technology a totally new approach to its application and use is required. The final stage is when it becomes clear that a new infrastructure is required in order to release the full potential of the new technology.

To keep a focus to the argument, the steam engine as a core technology of the first industrial revolution, was taken as an example to illustrate the working of these four stages of technological innovation in the nineteenth century. It will help us to do the same for the contemporary situation. However, the driving forces of the present revolution are many. Toffler, who has done much to popularize views about the techno-economic forces at work today, makes the contrast with previous periods when industries were primarily electro-mechanical. 'Instead, [the new industries] rose from accelerating breakthroughs in a mix of scientific disciplines that were rudimentary or even nonexistent as recently as twenty-five years ago - quantum electronics, information theory, molecular biology, oceanics, nucleonics, ecology and the space sciences'. It has been the developments in these areas that have led to the new industries of 'computers and data processing, aerospace, sophisticated petro-chemicals, semi-conducters, advanced communications and scores of others'. [3]

To bring this list up to date we could add the exploitation of new materials, particularly plastics and synthetic materials produced by biological rather than chemical routes, the use of ceramic material to replace metal for high temperature performance, optical fibres and lasers, and biotechnology. Out of this impressive list we shall concentrate on what is now probably the most pervasive technological development, information technology. The essence of information technology (IT), and hence its importance in contemporary economic and social change is that,

> To varying degrees all economic and social behaviour depends upon the manipulation of information - its production, processing, communication, storage and retrieval. It is the ability of IT to integrate these functions at low cost that makes it potentially such a powerful tool. The future developments of IT ... are very dependent upon the remarkable changes resulting from research in computing, telecommunications and related industries including: software engineering, intelligent knowledge based systems, human/machine interaction and opto-electronics. [4]

IT can therefore lay claim to being a core technology in the way in which steam power was identified in the first industrial revolution. The role of one particular semi-conducter device, a transducer, provides a way of describing the potential of information technology in a more down to earth style. For a transducer is a product/process which brings together three distinctive elements. The first is a sensor. The second is a built in logical process designed to react to that sensor. Third, there is an actuator which is activated in a certain way according to the reading of the sensor. A simple illustration of a transducer is that which can be found in any modern automobile for monitoring the temperature of the radiator water. There is a sensor which records the temperature of the water. The logic unit compares that temperature with a range of temperatures to see whether the reading falls outside a prescribed range. If it is higher than the maximum of the range it will then activate a light to come on the dashboard to warn the driver.

Although this is a simple and familiar example, a few moments reflection will reveal that it illustrates vividly the ways in which information technology now pervades so much of our daily activities. The same process is incorporated in the methods by which we access banking and financial services, where the drawing of cash from a cash dispenser involves precisely the same logic of sensing the customer's card, checking that there is money available in the account and then activating the dispenser to provide the cash. Similar logical sequences are regularly performed, for example, in a travel agents for booking holidays, or in a hardware store for checking what is in stock.

The spreading use of information technology in these ways, illustrated by this example of the use of transducers, has been made possible by the ability to make such semi-conducter devices small, fast working and cheap. The potential use still remains vast and many areas of economic activity currently relying on human intervention have yet to be exploited.

A pertinent (for economists) example related to information technology of the four stages whereby new technologies are introduced is the use of computers in economic forecasting. Prior to the introduction of computers, most economic forecasting models, such as those used in the British Treasury for medium term forecasting, were based on simple, mathematical models using first order equations largely to minimize the labour associated with solving equations by use of electro-mechanical computing machines. Even so they were still difficult to solve. As a result, typically, an annual forecasting exercise would probably take two or three months for the process of constructing and solving the model, whilst the forecast, given the time constraints, only permitted the exploration of one set of assumptions with one outcome.

When the first computers came along they were seen immediately as a way of eliminating the tedium involved in calculating the models. The process of calculation was speeded up enormously and immediately made it possible to look at more variants, alternative sets of assumptions and alternative outcomes within the same timetable. Thus, the first two stages of the introduction of new technology, i.e. the computer, were seen as a way of doing the same job that desk calculating machines had done previously, and then doing marginal new things, for example, by allowing more variants and different assumptions to be explored.

The third stage, which permitted a radical new approach to systems of forecasting, occurred when it was realised that the greater computing power available would enable, by iterative techniques, second or higher order equation systems to be solved. The constraints of calculation had hitherto meant that most economic relationships, for example, that between wages and unemployment (the Phillips curve), had to be formulated as first order equations. All mathematical models are, of course, only formal representations of cause and effect in the real world, but it is more likely that the inter-relationships of the real world will be better represented by higher rather than lower order equations. The ability to remove this constraint represented a radical improvement in economic forecasting methods. This advance led on to the creation of simulation models of various sorts and to complex economic forecasting models providing results of simulations as rapidly and as frequently as could be desired. Taken together with the increased ability to display results using visual display units, a total transformation in the methods of economic forecasting resulted. Finally, to underline the need for a new infrastructure to exploit the new technology completely, it became clear that the arrival of computers also opened up the possibility of linking the forecasting models directly into many newly computerized data bases, such as the censuses of production and distribution, and Customs and Excise data on imports and exports.

Other examples where information technology has resulted in major changes include the fields of communications, information data bases and computers. Initially the developments in each of these areas were seen as separate, but they have quickly become inter-related and progress in one can only follow from progress in the other. Whilst there is no reason to be pessimistic about the ability of the industries concerned to reconcile and integrate all these various trends, it is worth noting that one of the problems relating to infrastructure is not technological but political. The development of computers, and to a large extent now, information data bases, has been driven largely by the private sector. On the other hand, the communication networks in various countries are invariably under the control of national governments for national security reasons. This conflict of interest, so to speak, will need to be resolved. As the trend now is to integrate developments in the three areas of communications, information data bases and computer technology in order to serve an integrated market, a common information highway, it is impossible to think in terms of national systems. A global economy is already operating in these fields. Whilst it is understandable why some countries are concerned since they risk losing control over a vital new technology likely to have important repercussions for their economies, the creation of an international information highway is an excellent example of the need to get an infrastructure into position if the full benefits of new technologies are to be exploited.

Another example relates to the use of computers in manufacturing. The contrast between the processes now used, based on computer-aided design and computer-aided manufacturing, and those that prevailed ten or twenty years ago, is quite remarkable. The laborious task of redesigning tools, to provide the basis for a new production line (varying the characteristics of the product), could take many weeks, or even months. This delay has now been virtually eliminated by modern computer aided design and manufacturing systems which have given manufacturing processes the ability to respond to continuous and detailed modifications of product specifications.

A third example is in the use of information technology in the office. Again, what were originally seen as quite discreet tasks, such as typing, filing and communication by telephone are now all enveloped in integrated office systems based on word-processing, faxes and telecommunications networks which are constantly being improved and added to. Computer based operations are also spreading into the home, enabling the domestic television unit to be used for shopping, searching the lists of products available in stores, and for carrying out banking transactions.

Finally, the technology is being exploited in a range of other ways in and around the home particularly to carry out monitoring functions. These include security systems and systems able to monitor the smooth functioning of the various service elements in the house, such as gas, electricity, heating and air-conditioning as well as the working of the increasing stock of electronically driven equipment in the kitchen. There is also a potential for monitoring the conditions in the garden to measure, for example, whether the soil has a high enough moisture content.

These examples could be added to almost indefinitely, but they leave no doubt that information technology (together with the potential impact of the other technologies already mentioned) is producing a period of change, as rapid and revolutionary as that which took place in the first industrial revolution. As with the core technology of the industrial revolution, the rate of diffusion of the new technologies is the key feature of the revolution. It affects industrial processes, changes occupational patterns, and reshapes patterns of demand. [5]

The growing use of information technology is often linked with another feature of the modern economy, the growth in the service sector. The data in Table 3.2 showed how employment in the service sector in the economy is rapidly increasing; a trend which has been going on for many decades now. In one sense the trend is to be expected. The familiar concept of the Engels curve, which relates the proportion spent on various items of consumption to levels of income, suggests there is a distinction to be made between spending on basics, where the proportion falls as income grows, and on luxuries, where the proportion rises as income grows. Many service activities fall into the latter case and thus offer a part explanation of what is happening.

However, before going on to discuss the consequences of the trends in more detail, it is worth remembering that there is a close interaction between the demand for services and the demand for goods. To take an obvious case, quite often there is a complementary demand between a service and a good.

When we buy many of the consumer goods based on information technology, such as television sets, smart kitchen equipment, personal computers, etc., they often carry with them a commitment to service the equipment. The purchase generates a demand for a service as well as the purchase of the good. Equally, whenever a service engineer goes to do his job, he will carry with him goods, such as tools, materials and parts, essential to carrying out the servicing function. A more important example is the demand for capital goods (the output of computer manufacturing industries) generated as the use of computers and information systems replaces old style systems.

There is also the vexed question of boundary changes, that may affect the way statistics are collected but blur the distinction between goods and services. For example, until recently most major manufacturing companies would expect to have their own research and development (R & D) units. As such, scientists employed in these divisions, as employees of a manufacturing unit, would be classified as working in the goods sector. Recently many companies have shed these functions, reducing the staff and employing outside contractors for research and development (often employing the same people shed by the manufacturers). However, as these R & D units are classified in the service sector, there is an apparent reduction in the employment in the goods sector, matched by an increase in the employment in the service sector, although in many respects nothing has really changed.

It is also a fact that the balance between using goods or services in the home has changed over time. To take an obvious example, in Victorian days, most middle class families employed servants to do the household chores. The practice decreased as the tasks were taken over by the twentieth century middle class housewife herself with a little help from goods, such as vacuum cleaners. In the present day the increased sophistication of white goods equipment for use in the home requiring servicing reintroduces the servant (the service engineer).

Also, as a word of comfort, for those who fear the disappearance of the goods making sector, it needs to be remembered that despite the inclination of western economies towards service industries, there is still a large potential demand for goods in the rest of the world, that will need to be met from a manufacturing base somewhere in the world. If western manufacturers fail to capture or hold on to this market and the goods are supplied by other manufacturing centres, located, for example, in the Far East, our manufacturing sector will continue to shrink, but it is for competitive reasons and not because of the Engels curve or any other trend at work.

If the new service economy is to be the norm, it is desirable to be sure that the term 'service' is properly understood in its economic sense. [6] As a starting point it helps to think of a service as an act that changes the condition of a good or a person, and to make the distinction between services to goods and services to persons. Examples of services to goods are the transportation of goods from one place to another, postal services taking letters from one place to another, repairs and household cleaning. The feature of these services is that they are often akin to producing the good. For example, those who repair cars require virtually the same skills as those who make the cars, and the replacements, which are a part of the repair, are the same units that are put in the car in the first place.

Examples of services to persons are travel services, beauty services, medical treatment, and education. As these all affect the condition of a person (their sense of well-being possibly), they lie closer to higher (non-material) purposes of human activity, and often raise conceptual issues or boundary problems in their measurement. Often the services could be provided by individuals themselves or by friends, and thus not enter the market at all, or in some cases their benefits are almost indistinguishable from the benefits that come, for example, from the acts of thinking, learning, sleeping, etc.

One possible conclusion to draw from these reflections is that all labour activity is in a sense a service, whether it is engaged in a process, the end of which is the production of a good, or whether it remains as a service. For that reason it may be desirable to play down the distinction between goods and services.

Another consequence, prompted by the emergence of increasing levels of employment in service activity is to question the relevance of much of economic theory to these new circumstances. The bulk of economic theory has to date been built around an analysis of the goods sector. For example, business cycle analysis relies heavily on the inventory (stock) cycle, whereas the concept of an inventory, by definition, does not apply to services. The theoretical structure tracking the consequences of exchange rate changes, and the econometric work to quantify it, have been built around the response of imports and exports of goods to such changes. The analysis of labour market responses has also been in terms of the reactions to employment, unemployment and the time lags involved in the shedding of labour within the goods sector. The production function itself has always so far been fitted to data for the goods sector.

It is these analyses which have contributed to an understanding of the dynamics of the economic system in terms of full employment, inflation, balance of payments equilibrium and optimal growth. It has to be asked, therefore, whether there need to be service equivalents of this form of analysis? Is there any reason to believe that the business cycle will be less extreme now that the service sector is so large? What are the production function equivalents for the service sector? What are the reactions between changes in demand and investment? How does the labour market in the service sector react to changes in business conditions? What are the international trading patterns and responses for the service sector?

There are three other aspects of the service sector that are worth mentioning. The first is the growing importance of women employed in the service sector. For example, the proportion of women in employment in various British industries, comparing 1985 and 1994, was as follows:

Table 3.3
Proportion of women employed - by industry (%)

	1985	1994
Agriculture, forestry and fishing	26.4	26.4
Energy and water supply industries	12.0	23.0
Manufacturing	29.3	30.0
Construction	12.0	15.9
Distribution	54.4	54.6
Transport and communication	20.4	24.2
Banking etc.	49.2	50.3
Other (including public services)	62.9	68.0
All industry	44.3	49.4

The principal industries employing women are banking and finance, distribution, and other including public services, but it will be seen that the proportion is rising in almost all industries. This reflects the rising economic activity rate for women which grew from 45.5 per cent in 1981 to 53.2 per cent in 1990. Britain has the highest female activity rate of all western countries apart from Denmark.

The second point to note is the preponderance of part-time workers in the service sector. Figures for Great Britain are as follows:

Table 3.4
Percentage of available jobs for men and women taken by part-time employees

	1985	1994
Agriculture, forestry & fishing	9.5	21.8
Energy & water supply industries	9.8	5.0
Manufacturing	6.1	7.6
Construction	5.1	8.2
Distribution	31.4	44.3
Transport and communication	4.3	10.4
Banking, etc	12.3	17.3
Other (including public services)	31.1	41.9
All industry	19.0	28.4

As would be expected, the part-time rate is highest where female employment is high but the trend is steadily upward.

The general perception of employment in the service sector is that it is a lowly skilled sector, since the types of jobs that readily come to mind are jobs such as shop assistants and domestic servants. However, the service sector includes a large range of skilled professional services. Although figures for western economies generally are not readily available, the following figures for Japan relating to the early 1980s, derived from OECD sources, are probably generally indicative.

Table 3.5
Proportion of graduates in employment in industry (%)

Distribution	16.5
Finance	30.9
Transport	11.8
Public utilities	20.0
Professional services	35.7
Public administration	29.6
Total services	24.4
Total goods and services	17.9

This statistical review of current employment data suggests a number of conclusions. First, the present period is showing shifts in employment patterns of a scale similar to those of the nineteenth century. Second, within the overall economy, the service sector is continuing to grow in terms of employment with a trend towards more female employment and towards more part-timers. Third, the skill levels throughout the service sector are higher than perceived and can be expected to rise further.

The evidence of the Engels curve suggests that to a large extent demand patterns in an advanced western economy are driven by consumer spending habits, reflecting changing lifestyles. At present, in Britain, private consumption represents roughly fifty per cent of total final expenditure, whilst private and public consumption of goods and services together account for over sixty-five per cent. The need, therefore, to understand why people consume goods and services has always been very important. It is an area of demand of considerable concern to economists.

However, it also raises questions going beyond the relatively narrow discipline of the economist. Can a sense of direction be discerned in our consumption patterns? Is there a sense of purpose? Or is it all just aimless spending at the whims of advertisers? The immediate answer to these questions has to be that what we consume does matter very much. It matters because of its contribution to our sense of well-being. It also matters because consumers' spending raises a number of moral and ethical issues because spending transactions reflect relationships between people and are an integral part of the social network (Veblen's [7] work on conspicuous consumption provides many examples of the importance of consumption in society). All of us need, from time to time, to address the precise role played by material possessions in the pursuit of happiness. Economists do society no favours by accepting this contemporary compulsion to consume uncritically.

The economist's approach to consumer behaviour is broadly as follows. The conventional explanatory model for consumer behaviour consists of a set of explanatory variables, usually income, relative prices and wealth. The structure of the model normally assumes the law of diminishing marginal utility - the accepted underlying hypothesis of market capitalism. The model further assumes perfect markets and perfect knowledge, and the independence of behaviour of individual decision-makers. It accepts tastes as given. The model is assumed to be amoral and assumes that acts of consumption can be analyzed separately from acts of production. An analytical presentation of this form can be found in most economic text books. Theoretical formulations of this type have also provided the basis for fitting consumption functions to data and are invariably incorporated in aggregate forecasting models used commonly today. It is fair to say that such models perform very well for short term forecasting tasks, such as those used to support annual budget exercises.

Some economists have expressed concern about the narrow conceptual basis of these models. The assumption of independence of behaviour was challenged by Dusenberry [8] as early as 1949, whose subsequent work represented the first beginnings towards constructing a more social model for explaining consumer behaviour. Also, Dusenberry challenged the assumption that consumer decisions are reversible and pointed out that cultural pressures, and the desire to emulate social peers, made it very hard for people to reverse a particular standard of living once it had been acquired, even if they were to fall on hard times. Further research enlarging the analysis of consumer behaviour came from the work of Friedman [9], and in particular through his emphasis on the distinction between permanent income and windfall income. His subsequent analysis of the factors affecting the disposition of permanent income for short and long term patterns of consumer behaviour has enabled economists to take a wider view of consumer behaviour.

However, the conventional approach (even when enlarged) needs to be complemented in at least two respects. First, to consider what practical changes are now occurring in the way in which people behave as consumers in the market place, as a result of the impact of new technologies, particularly information technology. Second, to look at the explanations of consumer behaviour that other social science disciplines can offer.

The arrival of processes and goods based on information technology is affecting consumer patterns in many ways. A variety of labour-saving equipment is now available in the home, most of which are now smart in the sense that they are programmed electronically; to carry out sequences when unattended, to start and stop at preset times. The availability of these products has coincided with, if not encouraged, the trends towards greater female participation in the workforce. There is now a high premium on equipment which reduces labour in the home. Women prefer to work and buy equipment such as washing machines with their income in order to reduce household chores which in turn makes working bearable and practicable.

Also, although only a middle class phenomenon as yet, families are finding that the home can be used as the work base, both for husbands and wives. Access to customers or to parent organizations can be almost remote via telecommunications, the computer and/or television screen. Many salesmen now work largely from home (adequately linked by computer to their base) with only an occasional need to report in person to senior management. Husbands and wives can allocate the various tasks of income earning, housekeeping and child raising between them. The children themselves can use the work station as a means of improving their education, although that advantage is but one element in a complicated balance sheet of the good and bad effects of the new working styles in family life.

The many changes taking place in the transportation system provide another example. The consequences of the switch in consumer preference from rail transport to the motor car are well known, but, with congestion increasing rapidly and with a better understanding of the environmental costs of the private car, interest is now reverting to communal transit systems. Passenger transport is, incidentally, an interesting example of the complexity of service/goods inter-relationship. Consumers have moved from purchasing a unit of transportation, i.e. a rail ticket, to owning a durable good, i.e. a car, in order to provide the same service individually. Information technology has much to offer in the field of ground transportation. Computer guided systems will offer totally flexible systems enabling buses to call at the home on demand. They will offer greater speed and safety control on motor routes, with route finding facilities within the car itself. A car driven by an automatic pilot is no longer a fantasy.

Many of the changes in the patterns of consumer spending are related to changing patterns of work. The example of working wives, associated with a rapid increase in equipment in the home, has already been mentioned. Both partners working has also affected child care and patterns of eating, particularly an increased demand for convenience foods either in the home, centred around the television, or outside, to dine, to graze or to eat 'on the hoof'.

The provision of shopping facilities for consumers has also undergone rapid changes, largely related to the rise in car ownership, providing freedom to drive outside towns to enjoy the convenience of large shopping complexes. The effect on other shops, in villages, on street corners within cities and city centres are familiar to all concerned with inner city and urban renewal and preservation. There are many developments to come. There is already a significant increase in catalogue shopping and in the use of, for example, Yellow Pages, to find sources for consumer purchases. Already it is possible to shop 'at home' via televized catalogues which provide access to what is on offer at the various stores in the local shopping precincts. Limits to the expansion of this type of shopping is simply the 'smartness' of the electronic devices in the televideo system and the availability of a large enough data base to allow full comparison across stores (an infrastructure problem).

New lifestyles have created a significant increase in those transactions which lie on the margin of the definition of the market economy. It may take the form of barter between two professionals or skilled tradesmen. It may be firmly in the 'black economy' to avoid tax. It may be in those activities that are genuinely marginal between personal activity providing a service outside the market or providing the same service within the market. The choice between hairdressing at home or at a salon is a typical example. Or, to take another example, how do we distinguish between that part of children's education, which is done formally through attendance at colleges, distance learning in the home and that which can take place in the family using videos, partly for pleasure, partly for educational purposes?

These trends indicate that patterns of consumer spending are undergoing revolutionary changes although they still underline the dominance of acts of consumption in driving economic activity.

Reference has already been made to the weakness of the relatively narrow and self-limiting approach of economists to an explanation of why people consume goods and services. If the economist's approach is limited, it is worth examining what other social science disciplines have to offer by way of explanation of consumer behaviour. By doing so, differences of approach leading to equally distinct conclusions are opened up.

Economists concentrate on the desires and choices of the individual assuming that they are formulated independently of others. Sociologists, however, assume that it is the social nexus which is important and definitive in determining spending habits. It is the act of exchange, or in the sharing of things in return for similar gifts received in the past, that really matters. The sociological hypothesis immediately exposes the weakness in the economist's approach in that he does not attempt to integrate his model of consumer behaviour into a model of whole society. If the economist is to improve his model in this way he needs to replace the concept of the consumer as an individual, rational, agent with one that treats the consumer as a social animal. He also needs to take an interest in the use of the goods after purchase.

An authoritative writer in this field is the anthropologist/sociologist Mary Douglas. [10] What follows is a summary of her argument. First, it is impossible to think solely in terms of individuals. Social relations are the key. Consumption must therefore be seen as a part of a general use of signs and symbols in society at large. An underlying assumption behind this social activity is a belief that the social network has to make sense. The implications of being a member of a group, therefore, have to be recognized, i.e. the use of marking systems, the control of information and the various attempts to build up exclusive areas, keeping others not of the group outside those areas. Thus, since humans are social animals, consumption is part of culture.

It follows from this approach that goods can be regarded as markers and communicators. They say who people are and determine their sense of worth. 'Goods are endowed with value by agreement with fellow consumers'. [11] 'But at the same time it is apparent that the goods have another importance: they also make and maintain social relationships.' [12]

Important elements in the individual's life cycle are common determinants of consumption patterns, for example, birthday celebrations, weddings, funerals etc. Much is made of the exclusive nature of groups that control entry into them, determine patterns of celebration and the type and location of homes.

This sociological approach offers advantages in understanding why people consume goods which are not available from the economist's approach. They stress the sense of worth created by possession and use of consumer goods and services and their use as a means of communicating within a social network. They also link consumer behaviour to the activities (of the same consumers) in the production process, since status and role at work fulfil the same function as markers of who people think they are.

The approach leads to some important practical conclusions. For example, it gives additional insight into the choice between goods and services. It raises the social implications of some elements in consumer spending, such as spending on transport or on telecommunications, both of which are key elements in binding together a social network. Douglas' approach also explains why unemployment is so devastating to people, since unemployment attacks not only their purchasing power but their position in society and their sense of worth.

As Mary Douglas concludes,

> Each free individual is responsible for the exclusiveness of his home, the allocation of his free time and his acts of hospitality. The moralists who indignantly condemn over-consumption will eventually have to answer for whom they themselves do not invite to their table, how they wish their daughters to marry, where their old friends are today with whom they started out in their youth. [13]

> Goods are neutral, their uses are social; they can be used as fences or bridges. [14]

It is indeed a formidable alternative view of consumer behaviour to put against the economist's. Economic theories may be reasonably adequate for certain purposes, but they only address a small part of the problem. Wider views are needed to understand trends in consumption patterns in the long term, and to make sense of what is currently happening. Douglas' view will serve to remind us of the sovereignty of the consumer; it will also serve to remind us that consumption is a social activity which helps to explain want, not need, and places the economic model in the context of a wider social model.

One of the features of the first industrial revolution was the creation of large production units. The main cause of this development was the need to group people around the sources of power and machinery and to achieve economies of scale, the benefits that come by specialization within the sequences of production.

With the new computer aided technologies and the growing importance of service as distinct from goods industries, the importance of large scale units has rapidly diminished. The last thirty years have, in fact, witnessed a quite massive reduction in the size of industrial units and, increasingly, the location of these units away from the old industrial centres, since there is no longer any substantial advantage in being so grouped together.

It remains to be seen what criteria will ultimately dictate the location of industry in different parts of the country, or for that matter in different parts of the world. As has already been said, already most sales forces are able to operate out of home effectively with regular reporting via their computer network into headquarters leaving the necessity for meeting with senior management as an occasional requirement. Corporate structure is also being transformed with the development of core units within corporations, with many ancillary activities being undertaken by a workforce on self-employed or temporary terms. Whilst this new type of unit may appear efficient and appropriate to the new technologies, its introduction has the effect of concentrating controlling power of corporations into fewer hands than previously and results in a large part of the workforce no longer having the security of employment or the benefits of long term employment such as promotion or pensions.

The consequences of these new trends in industrial organization are considerable. Much of the existing urban transport network has been created to move large numbers of people each morning into the same place of work, and taking them home again in the evening. Now that this rhythm is a thing of the past it is no wonder that urban transport systems are finding it difficult to pay their way, and may ultimately have to be replaced totally.

The differing working patterns, particularly for semi- and skilled workers now made possible by new technologies, have already been referred to. We have noted that the relationship between husband and wife is changing. The role model of the husband in full-time work with a wife at home to care for him and a family is rapidly becoming a thing of the past. Increasingly the role of wage earner will be shared between husband and wife, partly by both of them working either full-time or part-time. An alternative will be for husband and wife to take it in turns to be the principal wage earner, either allowing the wife to have children or for the husband to change jobs, to gain further educational qualifications etc. Networking, in which people will work mainly out of their own home in dispersed units is already a viable form of working. The home has become an office for many people.

Although examples of these new work patterns already exist, it would be true to say they are principally to be found amongst the middle class. However, the main impact of the dramatic reduction in employment in manufacturing and the growth in service jobs has to be borne by the unskilled workers and in that sense, these trends will demand the most change from those least able to cope.

Studies of employment prospects [15] suggest that one of the problems created by current trends in occupational patterns is a substantial shift away from lower to upper groups, i.e. from unskilled lower socio-economic groups to skilled higher socio-economic groups. Since recruitment in jobs currently held by the higher groups tends to be self-sustaining, as indicated in Table 3.6, the result is that the lower groups not only have fewer jobs to look for, but increasing difficulty in breaking into the other expanding job areas.

Table 3.6
Inter-generational occupational change

Occupational Group	Fathers %	Selves %
Higher	13.0	25.0
Intermediate	33.0	30.0
Lower	54.0	45.0
Total	100.0	100.0

The higher group is defined as managers, professionally qualified workers and administration staff. The intermediate group covers clerical and routine non-manual occupation. The lower group covers manual occupations. The table measures the inter-generational shift up to 1972 by comparing the spread of occupations of men with those which their fathers had when the respondents in the sample were aged fourteen. The table shows that there was a switch in occupational patterns resulting in a doubling of the percentage in the higher group within one generation at the expense of the lower group. The prospects for manual workers and unskilled workers are worsening significantly. Moreover, data analyzed in respect of 1983 (see Table 3.7 below) showed that the bulk of long term unemployed is to be found in the lower group.

Table 3.7
Occupation and unemployment

Occupational groups	Proportion in population (1983) %	Proportion in long term unemployment %
Higher	34.0	4.0
Intermediate	27.0	21.0
Lower	40.0	75.0
Total	100.0	100.0

The new service economy, with its continuing momentum of technological change, is likely to continue to generate steady rates of economic growth and to continue to offer major opportunities for pursuing new and affluent lifestyles. Society is benefiting from new technologies resulting in a steady stream of new products and services. The exploitation of the new technologies is causing, in Britain, radical changes in industrial organization requiring substantial shifts in employment. These shifts have been reflected in a changing geographical pattern, from north to south, and in a changing industrial pattern from manufacturing to service industries. The extent of the change is reflected in the fact that employment in manufacturing fell from 7.3 million in 1977 to 4.3 million in 1994. Working patterns have also been revolutionized. The disappearance of large scale industrial units has destroyed the old link between population and workplace and introduced changes in contractual working; from largely employed with reasonably assured terms and length of employment, to more self-employment, part-time working and short term contracts.

The ways in which institutions, both those closely linked to economic activity and others in society at large, have been forced to respond to present changes will be discussed in more detail in later chapters. However, the extent of change is not in dispute. The new economy is obliging all to come to terms with it, including those concerned with the appropriate value basis of the various constituents of society.

Many are benefiting from economic change and from the many new opportunities. The winners are also full of confidence and seek to exploit their gains to the maximum by eliminating obstacles to further expansion. In contrast, there are many losers. The costly transition from rural to industrial of the eighteenth and nineteenth centuries is not being repeated but the devastating shift out of manufacturing is taking the same toll on large numbers of the labour force. There is also a widespread attack on many supportive and protective institutions, particularly those related to social welfare, that have been created in reaction to the previous onrush of economic change. Non-welfare institutions, such as the legal or educational systems, are also being challenged to adapt to new circumstances.

There is an added dimension today. Such is the flow of technology and the scale of economic activity that has been achieved, that the ability of the planet's resources to sustain these levels of activity and rates of growth is increasingly being brought into question. In this respect new and novel issues of concern have to be added to the agenda.

For instance, society is increasingly becoming aware of, and concerned with, the environmental impact of economic activity and affluent lifestyles. Current industrial systems create high levels of pollution. Examples abound; the effect of acid rain, of noise and water pollution, the risk in using nuclear materials. The balance between urban and country living is now precarious and the issue is focused, particularly in Britain and the rest of the European Union, on the future policy for agriculture. There is also concern about the conditions under which biotechnology and the techniques of genetic manipulation are being developed. Their application is not only in medicine but also in the development of food products and new materials. New techniques appear to be currently pressing against present ethical limits particularly in the field of embryo and human cell research where there is a natural human concern. The science is also making huge advances in animal and plant genetic manipulation where perhaps the same concern should, but is not, being displayed. Perhaps the point is being reached beyond which humankind ought not to go. Such concerns are already slowing down the rate of development and introduction of many new drugs.

It is not surprising that there is an element in public opinion which is reacting to these concerns by searching for modified or alternative lifestyles. The emergence of Green parties throughout Europe is one particular example of this phenomenon. Besides factors relating to environmental impact there is another issue which, whilst no longer in the forefront of public debate as it was twenty years ago, is still relevant; namely the risk that the world will soon run out of many conventional resources upon which the present motor of economic development runs. The economist has a simple (if naive) answer to these problems, which is that the price mechanism will respond to incipient shortage of resources. By increasing their price, the price mechanism will spur on the search for alternative new developments to provide replacements or substitutes for resources. In view of the global dimension of the emerging resource shortages, the economist's somewhat innocent approach looks increasingly inadequate. The 'green' alternative view of the need for stewardship, to respect nature and the use of global resources, by inhibiting rates of economic growth, is finding increasing support. A common feature in all these reactions to present levels and style of economic activity is that they are emerging from groups starting with a different value base to that which underpins economic activity.

Notes

1. Roll, Eric (1978), *A History of Economic Thought*, Fifth edition revised and enlarged, Faber and Faber Ltd., p. 295
2. Deane, Phyllis and Cole, W.D., op. cit.
3. Toffler, A., op. cit., p. 150
4. Long-Terms Perspectives Group (1985), *IT Futures*, National Economic Development Office, London
5. There are many sources for further detail covering the nature and consequences of this technological revolution. They include, besides Toffler, Alvin (1980), *The Third Wave*, Pan Books, London, Naisbitt, John (1984), *Megatrends*, Macdonald and Co., London, Gerschuny, J. and Miles, I. (1983), *The New Service Economy*, Francis Pinter, London, Gerschuny, J. (1978), *After Industrial Society*, McMillan, London.
6. For further study of this interesting phenomenon the reader is recommended to read the seminal article by Hill, Professor T.P., 'On Goods and Services', *Review of Income and Wealth*, December 1977
7. Veblen, T. (1925), *The Theory of the Leisure Class*, Allan and Unwin, London
8. Duesenberry, J.S. (1949), *Income, Saving and the Theory of consumer Behavior*, Harvard University Press, Cambridge, Mass.
9. Friedman, M. (1957), *A Theory of the Consumption Pattern*, Princeton University Press, Massachusetts
10. Douglas, Mary (with Isherwood, Baron) (1979), *The World of Goods*, Allen Lowe, London
11. Douglas, Mary (with Isherwood, Baron), ibid. p. 75
12. Douglas, Mary (with Isherwood, Baron), ibid. p. 60
13. Douglas, Mary (with Isherwood, Baron), ibid. pp. 11-12
14. Douglas, Mary (with Isherwood, Baron), ibid. p. 12
15. White, M., 'IT and the Changing Structure of Employment', research paper quoted in *IT Futures - It Can Work*, NEDC, London, p. 5.13

4 Society in flux

The previous two chapters have described two periods of rapid economic and social change driven by new technologies. Both illustrate the characteristics of an economy on the move; how the essential ingredients of economic development relate together.

A broad sequence can be identified normally starting with the process of scientific discovery. Modern society is characterized by the amount of resources it devotes to pure research - blue skies research as industrialists often describe it. Of course, the direction and priorities given to scientific effort at the universities or in related research institutes is influenced by the source of the funding (which, in some cases, such as defence and security, is substantial). This bias may be benign in the sense that it does not substantively obstruct pure scientific enquiry or infringe the universality and openness of rules governing scientific method.

Modern society, however, both supports and is supported by a vast and complex activity of scientific research which is ulterior in purpose. The activity is a complex and powerful element in society which it would be idle to presume can be easily and fully controlled - as can be seen in new areas such as biotechnology and genetic manipulation. Today's research activity is on an unprecedented scale and is the source of an onrushing flow of new discoveries, most of which are capable of being translated into new technologies.

The task of moving from scientific discovery into applicable technologies is important and complicated. Indeed the problems associated with application of new ideas are often crucial. The inability to get the Savery steam engine to raise water more than a height of a hundred feet eventually, and crucially, limited its use to watering ornamental gardens or pumping water out of mines. In a contemporary context, whilst the principles of the system based on semi-conducter devices to control fuel injection systems in automobile engines, are fully understood, the crucial task for motor manufacturers is to get such devices to work under the extreme conditions existing under a car bonnet (heat, vibration, dirt, etc.).

Even when discovery has been turned into technology, there remains the final stage of creating the economic conditions under which new technologies can be exploited profitably. These conditions are potentially many and they may include the need to overcome parallel technological problems which arise in setting up manufacturing processes to make the new products. Often new technologies demand a new form of manufacturing method or at the very least new approaches to minimizing costs of production. Often the introduction of new technologies requires an equally innovative enquiry into the market potential for the new product involving entrepreneurial judgement on the precise form of the product and its market. Overcoming these obstacles is a necessary economic condition which, if not met, can result in the failure of products, using the new technology, to find their way into the market. Given the importance of the various stages (described earlier) under which new technologies seek to enter and take over a market, the judgement to adopt a new technology and exploit new markets is a fine one. It is not surprising that many firms adopt a low risk strategy towards introducing new technologies, preferring to wait for others to be the first to test the market.

The nature of the inter-action between science, technology and economics is easily recognized in the modern world. But similar factors applied in the first industrial revolution, the differences being mainly of degree. Far more scientists than today were able to retain an overview of the whole of the developments taking place within the physical sciences. The scientific community itself was infinitely smaller and thus knew of each other's work. More importantly the major discoveries, and the problems involved in their application, were basic. The application of the steam engine, once the Watt engine was made reasonably effective, faced few problems in its introduction to different types of factory production. In the key industry of the early stage of the industrial revolution, textiles, the principal and only essential technological innovation, was the introduction of the spinning jenny. Improvements in weaving technologies were not required for another fifty years. [1]

The need for capital funds was not great in the early stages of the first industrial revolution - compared, for instance, with the cost of technological exploitation today. As a result, many small capitalist entrepreneurs were able to develop new technologies and exploit new market opportunities. As Hobsbawm points out, this form of industrial development would eventually pose serious problems for British industry by leading to

> the extremely decentralized and disintegrated business structure of the cotton industry, as indeed of most other British nineteenth century-industries ..... Such a form of business structure has the

advantage of flexibility and lends itself initially to rapid initial expansion, but at later stages of industrial development, when the technical and economic advantages of planning and integration are far greater, develops considerable rigidities and inefficiencies. [2]

One characteristic of the first industrial revolution which is emphatically not true now is that Britain was the first to apply and exploit science and technology economically. There was nothing to import from elsewhere. At the very least this ensured that, for many decades, the British entrepreneurial spirit was given full rein at the expense, perhaps, of more sober economic calculus.

In both examples we can see the nature of the interplay of forces that provided the dynamo for the movement of an economy over time. The question of which is the dominant force remains unanswered. In the early nineteenth century the impression given by the evidence is of a dynamic created by a few entrepreneurs pushing at the leading edges, not only of technology but also of forms of business organization, in the context of a rapidly developing home and world economy. The part played in the stimulus of these developments by growing consumer demand and expectation, has probably been under-emphasized by researchers, perhaps, through lack of data to verify hypotheses. However, the anecdotal evidence in Chapter Two provides sufficient grounds to support the argument that rising consumer power must have played an important part in stimulating technological innovation too. No-one would doubt the need to analyze the role of consumer demand in stimulating technological innovation in the late twentieth century. It can be argued that economic tastes and expectations are artificially stimulated by the production system itself, but that is not a satisfactory or sufficient explanation of the dynamism of the modern consumerist society - to which we will have to return later. What is clear judged by the rapid change in the nature of consumer and capital goods, is that the late twentieth century provides ample evidence of the dynamics of rapid technological change and of the radical consequences that follow.

The two case studies drawn from 1750-1850 and 1945-1995, confirm the radical nature of the changes that took place. The transformation that took place between 1750 and 1850 was in many respects a bitter and cruel episode that served to open up the possibility for even greater change in the century to come. It was a period exploited by many in the search of great gains at the expense of others who were harshly affected. By 1850 a form of industrial organization had been imposed right across the country which, if not new in form (since there were antecedents in the preceding century), was at least novel in the radical nature and extent to which it attacked and destroyed existing structures.

The list of new things is large. A vastly different pattern of consumption, a transformed set of relationships with the rest of the world, based on the growth of empire, a novel way of doing business led by a new type of entrepreneur encouraged by the formation of the limited liability company and oppressive new working conditions to which an uncomprehending workforce had to conform.

As Hobsbawm points out [3] there were many who were able to turn these radical changes to profit. Unusually in Europe the landed gentry saw the merit of swimming with the tide. They were not bound as their continental counterparts were to a feudal way of creating wealth from their land. They saw what was to be gained by mining their own land, by selling land for property or railway development or in engaging in entrepreneurial activity themselves and exploiting their own resources. Nor were they too resistant to seeing their labour force shrink - a result of the push of continued enclosure movements and the pull of the new industrial regions. Not surprisingly the existing hierarchy of dependent professions, of law, academia, medicine, the church and the armed forces all fed comfortably off this new level of prosperity.

More importantly the period saw the emergence of a new middle class whose activities lay at the heart of entrepreneurial activity. At first, the successful used their wealth to buy into existing social structures (to ape the aristocracy), but their numbers soon made it impossible for all to follow that traditional route. Instead they created their own social structure and social practices reflected in what we now term Victorian England (the consequences and characteristics of which will be pursued later).

Much has been written of the hardship of those who became the working class in this new regime. The numbers involved were vast. The extent of change imposed on them as they moved from an agricultural to an industrial working culture was huge. As Hobsbawm remarks, 'the measurement of life [was] not in seasons ('Michaelmas term' or 'Lent term) or even in weeks and days, but in minutes'. [4] Life in the familiar (even if often harsh in its own way) supportive system of the parish community was exchanged for a life in newly created industrial cities in which, amongst all the bad consequences, the break in social contact between better and worse off (anomie) was full and complete. In the long run the creation of this anomie was important for the country's economic, political and social development, but the immediate consequences were devastating in terms of human suffering. Ample evidence can be found of the horrors of the working conditions of the time in the exposures provided by successive Royal Commissions (into child labour, for example, in 1832).

The period since 1945 has been a period in which the flow of new technologies has continued with increasing impact. The British economy in 1945 remained distorted by the trading nexus of the British empire. That network was largely destroyed in the ensuing fifty years. It has been replaced by new links with the other members of the European Union, requiring a different product and market orientation for industry. The economy has moved from a heavy dependence on the manufacture of goods to essentially a service economy resulting in a significant displacement in terms of jobs, shifting from the goods to the service sectors. A particular feature of this displacement, which is still taking place, has been the move from non-information technology jobs to information technology based jobs. However this switch is now being followed, with gaining momentum, by the more threatening replacement of information technology based jobs by information technology capital.

This switch from the imperially based economy of 1945 (the population of the British Empire in 1947 even after India's independence, was still around one hundred million) to an economy reflecting the social democratic ideology of the 1950s and 1960s, through to the 1990s enterprise economy and culture, has been as dramatic as any fifty year period in the nineteenth century.

Those who have gained from it have exploited the new technologies and new sources of energy (particularly oil), the service industries, particularly financial services and the professions. Reflecting the growth of a consumer society, there has also been considerable job creation in the less skilled services area. This expansion has also brought about a growth in jobs attracting different working conditions; part-time, female oriented, non-unionized and lacking the security of pensions and other traditional benefits. The structure of companies has also changed to reflect the new enterprise culture. Workforces have been adapted into fast and slow stream personnel, the former being highly prized by management and well looked after, the latter treated more cavalierly and put at arms length and on a casual basis as far as possible. Whilst lacking the dramatic physically harsh working conditions of Victorian times, these changes in their own way have had an equally powerful impact on working practices for the mass of people.

As Table 3.2 in Chapter Three showed, the area where the post-war changes have caused most pain has been in the traditional industries of coal and heavy engineering (both associated with imperial trading patterns), steel making, rail transport and many other parts of manufacturing.

The cost, of course, is not to be measured solely in terms of lost jobs. It is also reflected in lost skills and in many instances in an inability or a lack of opportunity, to learn new ones. The shift of labour from one group of industries either into others, or into no job at all, and the distress it causes generally to the bulk of the working population, is a reminder that in a modern economy a job means more than a way of earning a livelihood. It represents position and status and is a generator of self-esteem and worth.

The economic and social effects created by the introduction of a new technology do not occur evenly. In the first two stages of its introduction there is neither a general awareness of the opportunities that will follow from radical change nor a realization of the threat that the new technologies are inevitably going to pose. The first stage of the introduction of the steam engine posed no new problems to existing industries or methods of working. It was seen simply as an alternative way of creating water power, by raising water from one level to another and letting the force of gravity generate power. Nor did the second stage, as steam was introduced into other areas where a more efficient pumping system was welcomed, i.e. in mining. Nor at these stages did the switch to steam engines raise difficult issues of industrial production or organization.

Similarly, in the modern information technology era, the first application of computers as a more efficient and speedier form of carrying out calculations presented no threat to industrial organization other than perhaps to the makers of electric calculating machines. Even the second stage, when the use of computers went wider and became more ambitious, the application possibilities were still within the orthodoxy of existing technologies and ways of applying them. This was true in the critical area of the use of information technologies into the office.

The real battle between old and new technologies takes place in the third and fourth stages. In the case of steam the clash resulted from, first, the widespread introduction of steam powered engines into manufacturing, particularly of textiles, creating the need for factory working - to exploit the full capacity of the new machines. Second, it flowed from the application of the steam engine to locomotion, which completed the massive transformation of British industry. In the last fifty years products based on semi-conducters, at the heart of information technology, have been able to replicate many areas of (service) activities previously carried out with the application of the human brain. They have also benefited from the exploitation of potential created by telecommunication networks and data bases.

Undoubtedly, in any period of rapid change, these two final stages take a considerable time to implement. There are good reasons for this. One of them is the sheer scale of technical problems to be overcome in achieving the final stage. Infrastructures have to be created - which by definition is a considerable undertaking. In the case of the steam engine, before its use for locomotion, when penultimately it was introduced as a power source for industrial machinery, factories had to be built, machinery invented or adapted to maximize the advantages of being linked to a steam source of power. These were the direct needs before the new source could be fully utilized. In addition there were the new consequential problems, of managing a large workforce unaccustomed to working within the disciplines set by the machines and the need to acquire and then teach new skills for managing factories and maintaining discipline. Infrastructure problems also included the need to cope with necessary ancillary changes, such as creating new housing close to the factory or teaching new working skills.

Finally, the use of steam for locomotion created similar infrastructural problems, but of an immensely higher degree. Putting the railway network into place, over a period of ten to fifteen years, was an economic revolution in itself and could hardly have happened without the availability of surplus investible funds at the time. The consequential increase in mobility created by the rail network affected industry dramatically - altering the location of existing industries, creating new ones and improving freight transport costs. It also altered consumer perceptions of lifestyle especially the ability to travel, both in the U.K. and abroad. But these processes took time and money.

The equivalent impact in the last fifty years has been the convergence of a number of differently conceived streams of development, all based on information technology. These include developments in computers, the introduction of information technology into the office, its use in manufacturing design and methods, in the home for controlling white goods and driving monitoring systems and in developing new home telecommunications networks. These streams are now merging with a radical impact on industrial organization and on consumer patterns. To get to this stage a substantial investment in information technology has been required, both to improve the technology itself and also to finance the necessary infrastructure. In the telecommunications field the task has also involved investment in space technology (including satellites), which has created a highly sophisticated telecommunications system, the information super-highway, that knows no national boundaries. When the then President Reagan observed in 1989 that the Soviet hegemony had collapsed, through its inability to defend its frontiers (physical and ethereal) against the invasion of western consumerism (and its underlying political and social culture), he was referring to this internationalization of information systems and indulging in only a moderate degree of hyperbole.

What the final form of the economic impact of the present technological revolution is likely to take, is yet to appear, but it is already clear that the range of consumer goods and services on offer is being transformed at the expense of major changes in working conditions and lifestyle.

The fact is that the final stages, at the end of which new technologies command the battle field, are inevitably time consuming and require much investment. It is a slow if inexorable process. Even without resistance from older technologies it is not surprising that new technological breakthroughs run into trouble. There will always be a degree of uncertainty attached to ventures that seek to break new ground - especially when the nature of that new ground is unclear. So new technologies cannot expect an easy ride. Many obstacles also stand in the way that are not all technological or financial. There is deep resistance to overcome. The new technologies and the fledgling industries seeking to exploit them, are up against well entrenched industries that are not likely to yield ground quickly or graciously.

Most important are the vested interests of the forms of production built around existing technologies. Since they are already well established, existing industries have the advantage of experience and well tried methods of operation. At the first sniff of a threat, therefore, a powerful defensive reaction can be expected which can take many forms. It may be directly commercial, in the form of an intensification of marketing and selling, or a minor improvement to the products made with existing technology, to meet the perceived threat of the new technologies. It may be a subtler, or more insidious, form of dissuading those who are essential to an effective launch of new technologies, for example, bankers etc., to think twice before supporting the newcomers.

In applying their opposition, existing companies can also rely on a number of groups sharing their own vested interests in things as they are. Certainly the workforce, with its investment in a certain range of skills, will react in a conservative way to new challenges. Understandably so, since lacking the ability to take any initiative in support of new technologies, they can but contemplate their arrival, and the threat they pose, with fear. But the defensive mechanism will be more deeply entrenched than this. The status quo will be able to call on, influential, elements in the nexus of services and professions created to support manufacturing industry. All, but an intrepid few, will begin by wanting to hold onto a status quo that gives them their livelihood and which would appear as much threatened as the old technologies themselves.

We can add the political system itself to this list. At local level, politicians will also react defensively if they see local manufacturers, using existing technologies, threatened - for there is no reason why the new technologies and the process of exploiting them, should set themselves up in the same locale as the old or employ the same number of people or people with the same type of skills. At national level the political breezes will waft along an even stronger smell of threat as lobbyists for existing industries bend politicians' ears and persuade them to perceive the threats as potential lost votes.

Finally, there will be the lynchpin of all conservative reactions - the inertia enshrined in existing structures and institutions. The power of this inertia cannot be underestimated. The task of overcoming it is usually the supreme challenge.

As a result of all these factors the burden of inserting new technologies into the economy, at least in the early stages, rests almost wholly on an initial small group of believers. The response of others may be so slow and insensitive, that many years or decades are required before the breakthrough fully occurs.

Market economics claims to offer a speedy resolution of one of the principal elements in economic activity - the balancing of supply and demand - by setting a clearing price. Much economic analysis is built around this fundamental proposition. It is the foundation for modern micro economics and in some key respects, such as international money and stock markets, at macro economic level too. But we deceive ourselves if we elevate this aspect of economic activity to the sole or even dominant role in the complicated process that goes on around us under the general term of economic activity. The foregoing sections have discussed the difficulties facing new technologies in trying to break through, and the resistance that is inevitably built up by interests vested in existing technologies. They argue that a major part is played in the way in which economies develop, by the group of institutions and structures involved in the process. Many of them are deeply entrenched in society and are therefore resistant to change and slow to respond. Yet their response is often a key to the successful introduction of change. Moreover, the key objective, freeing an economy to move on in new ways based on new technologies, invariably involves tackling the role of these institutions. Since they are large and slow to change the term 'heavy structures' has been used to describe them. The term originated with the eminent French economist, Claude Gruson, who described them as follows. [5]

> Heavy structures comprise many different elements - physical, organizational, cultural - where the process of creating, modifying or suppressing is very slow and costly and which are very difficult, almost impossible, to modify once they have taken shape. In fact, a strategic plan (hopefully to achieve change) must be tested for its compatibility with the heavy structures upon which its execution depends; put another way, the plan will only be fully articulated if it contains realistic proposals for the adaptation or the removal of such heavy structures, if that be needed.

Such heavy structures are often the principal causes for the delay in the introduction of new technologies. When the breakthrough eventually comes, in the third and fourth stages of technological evolution, we see a rapid almost precipitate transformation of these heavy structures. The momentum swings from using the resources entrenched in existing structures, to resist change, to using them to enable change to occur as rapidly as possible. Sources of capital swing their allegiance from the old to the new technologies, the problems of inadequate infrastructure are addressed as rapidly as possible; before long anyone that can has jumped upon the bandwagon.

The period, when the new technologies are fully exploited, puts a premium on the maximum freedom to act and to change. In the process it begins to define sharply the winners and the losers. It clears away the debris of past practices and puts power and control into the hands of a new generation or those of the old able and willing to adapt. Of course, before long, the battle won, a process of consolidation sets in. The new technologies, and the associated remoulded economy, create then their own heavy structures, firmly exercising power and control and unconsciously preparing themselves for their own defensive battles yet to come against the next wave of upstart technologies.

Creative destruction indeed. But this process reflects the two main features that dominate consideration of an economy on the move. The first is the constancy of change. The second is the intensity of the battle, often at the deepest level, as new technologies struggle to emerge and overcome the status quo.

As far as the first element is concerned, it fully reflects the experience of history, of the development of what we now call western market capitalism. There has always been debate about when the transformation of the mercantilist/merchant capitalism into market capitalism began. The foregoing sections suggest that there is less and less profit in trying to fix such a date - even concentrating on the mid-eighteenth century is too blurred for comfort. The continuing evolution of the economy is best seen as a rich, continuous and evolving process with no beginnings and apparently no endings.

The second element has also to be grasped more fully than we have done so far. Schumpeter's term relates to the economic process in its deepest sense. But it is appropriate to consider all heavy structures, economic and social, in the same way, since no-one can deny their fundamental relevance to the process of economic change.

Once the area of interest is widened it becomes clear that the list of those elements party to the process of developing the economy widens, and the wider institutions and structures of society have to be brought into the reference frame. When we considered the impact of changes it was necessary to invoke a wider social dimension in listing winners and losers. So, too, in exploiting the significance of heavy structures it becomes necessary to introduce the wider social dimension. In other words, the battle ground that we need to define is society itself, if we are to deal adequately, both with the causes and consequences of new technologies and the role of institutions. Many 'heavy structures' are institutions in the conventional sense but others are more complicated to assess, such as the education system, prevailing philosophies and underlying value systems.

The debate about the relationship between economics and society at large, especially if approached from an economic perspective, cannot fail to recognize the dominating vantage position that economic activity occupies in society. This is an obvious point to make even in observing early economic systems, for the large mass of people, up to and including the beginnings of the first industrial revolution. The biblical view of work, promising nothing but pain and sweat in order to eke out no more than a basic subsistence, would have been a commonplace view for the majority of people for most of the time period encompassed by the twelfth to nineteenth centuries. It was not so for a privileged minority and there may have been times, for example in the late Elizabethan period, when more people than usual had the sense of well-being that comes from an assured surplus after basic needs have been met. But for most, most of the time, life was grim and survival was a preoccupation. The same was true if measured in the sheer amount of time devoted to achieving economic survival. It could be simply put. Apart from the high moments of fairs and feasts, survival activity took up all daylight hours. It is not surprising that some form of economic model, embracing wider organizational and institutional aspects and a set of values which determines behaviour dominated life in most of its aspects.

It is a harder task to argue why, given the present levels of affluence in a modern western capitalist system, there should be the same degree of domination today. It is true that working hours have been substantially reduced over the past century. There have also been substantial improvements in working conditions for the majority. The majority have shared in the fruits of economic growth in a way never before experienced. However, it is still true that society appears to be content to allow present economic activity, its organizations and its institutions, and its accompanying value sets, to dominate behaviour and to influence society's direction as much as ever.

Most of the gains of economic growth have been taken in the form of greater affluence. The use of that greater affluence has been, in the main, devoted to providing more and more consumer goods. We are correct to call the modern economy a consumerist society. Why this continues to be so and especially why society seems to have insatiable wants, will be looked at in detail later. It is a fact of immense importance. The drive for more and more material goods is a preoccupying concern for economic activity. It justifies acceptance of ways of doing things, of particular lifestyles, which, prima facie, conflict with logic or a sensible view of wider social goals. It leads to an acceptance of value systems which dictate action, or impose subservience, in a way that conflicts with wider perceptions of social behaviour.

It has also led to, and provided, a context in which people evaluate their own or their neighbour's social ranking - to use a crude term for the moment. We are what we consume; we are what we do. The consumerist society is the arena in which people define themselves and their relationship to others. The complicated process, involving not only the purchase of goods for individual satisfaction, but especially their use as a means to show others who we are, is the one that most people appear to accept as the main route to the pursuit of happiness.

The general acceptance that life should still be dominated by the processes of production and consumption, allows those controlling and dictating economic activity to occupy a position where they can dominate the ways in which such economic activity is carried out. Those in control set the rules against which behaviour and performance are judged; they are also given a great deal of room to manipulate social institutions, and influence many aspects of social policy, if they are perceived to have a bearing on the way in which economic activity is being carried on.

The dominance given to economics is more pervasive than simply the sense of power by which the way the economy is run, at both micro and macro level. It carries with it a momentum which creates a set of values that both determine and justify action. How those values, which support new forces of economic activity, emerge is as interesting and as important as a study of the new thrust of technology and economic changes itself. The adoption of drastically different value sets at the beginning of the industrial revolution is a classic example, which we shall need to look at in detail later. But the new values accompanying and supporting new technologies and their consequences for economic activity which have emerged in the past fifty years, represent an equally dramatic change.

It is clear that just as gains from economic change fall first to some, and then to many, in a technological revolution, there are many who are happy to accept the new values by which new economic activity is being justified. It can hardly be otherwise. The new conquering armies need a belief that justifies their activity. That ultimate justification, i.e. a conviction that what they enjoy doing successfully, is justifiable in ethical, moral or religious terms, provides a powerful sense of life fulfilment. We must not seek to minimize this sense of missionary zeal.

In contrast, for a long time, many may well feel threatened by the practice of these new values. Other threats are great enough; such as the disappearance of industries using old technologies, the pain of adapting to the new, the harsh interregnum as the economy moves to another style and the failure of many ever to adapt. All these threats conspire to create a great sense of fear and uncertainty. The final straw may well be a realization that old values are being swept away. Values that had for a long time justified old systems and were consistent with wider social values. Their departure, perhaps the ultimate threat, leads to the great poignant cry of 'it's not fair'.

The extent to which these new values are adopted, to justify new economic activity, does not end there. It is quite possible, indeed it is highly probable, that they will be inconsistent for a considerable time with the value sets to be found in other parts of society. If they are, the reaction will depend mainly on the application of two criteria. First, if the clash of values reveals obstacles in society at large which impede the total victory of the new economic values, then the latter will seek to assert their supremacy and impose priorities - in the interest of new economic activity. Alternatively, such is the imperialistic nature of new, dominating value sets that they may lead to a root and branch reform of social institutions based on alternative value sets to re-establish them in a way consistent with the new economic value sets.

It is thus evident that the process of creative destruction and the necessary attack on heavy structures, whilst emanating principally from economic causes, will instigate and set in train similar forces in the wider social context. Where it does so, an issue of the relative importance of the economic and social models is intensely raised. Which is the dominant model? The narrow economic model or the wider social model? Logic says the latter but experience suggests the former. A logical, but less easily articulated argument, is that social change and values influence equally substantially the principles in which economic activity is based; that social values set the ultimate context in which economic activity has to be justified. This argument, important, indeed crucial, as it is, is less easily articulated; it is not as easy to talk about a social model, and its related set of values, as it is to talk about an economic model, and its related set of values. However, the case is there to be argued. Society as a whole provides an over-arching context in which to set economic activity. If the consequences of the latter or the practise of particular values in the latter, conflict with wider social aspirations it is clear which should yield. The plea, 'it's not fair' has to be heard and addressed against wider values than simply those supporting and serving the economic model.

Moreover, it always needs to be borne in mind that the past offers examples of many different types of economic systems based on different types of value sets. The mercantilist-post-mediaeval period is one such case. The small scale of geographically limited markets, serving towns of less than 50,000 inhabitants, created natural conditions for an intimate nexus of ethical rules governing business transactions, which included concepts such as a just price or a fair wage, and many conditions limiting the degree to which those in hard times could be exploited. Some will say, to hark back to an earlier period is no more than nostalgia. However, there are modern examples of economic activity, like the Japanese, just as effective in terms of economic success as European and North American ones, which build on different value sets. It also remains to be seen what system based on what values, for example, China or India will adopt as they participate more fully in western style economic activity.

Nevertheless, the presumption in western market capitalism appears to be that the economic model, to a large extent, drives and dominates peoples' lives to the point that the logic that says society will offer protection through the application of wider values, has to be abandoned.

When the way in which society itself changes over time is examined, it is quite clear that the same pressures for change are created, with the same resistance, the same battles and the same sequence of struggle. That struggle continues until new values, new ways of organizing society in terms of individual and community, or at least new balances are struck between the respective claims of these two pillars of social development, are put in place.

Many of these wider changes will have an impact, perhaps a crucial one, on the way economic activity itself is carried on. For instance, the creation and development of welfare institutions in the century ending immediately after the second world war, designed to mitigate the harshness of the workings of the market economy. Also, the difficulty the education sector has had over the past hundred years in finding an effective way of supporting Britain's economic development. There has also been the wide flux of beliefs, both religious and otherwise, which has sought to battle for the soul of economic activity.

As we come to the end of this chapter, the way in which the argument needs to proceed is, hopefully, becoming clearer. The two exemplary periods considered in detail describe what happens when an economy is on the move, inevitably under the stimulus of new technologies. The attempts of new technologies to dominate in terms of a battle in which success is at the expense of old technologies have been described. The creative force has its destructive side while the existence and role of heavy structures have been noted.

It has also been argued that movements in the economy cannot be viewed in isolation. Economic activity influences the lives of all since it is seen, not only in terms of its fruits but also in terms of participating in it, as a principal route towards life fulfilment, however that may be defined. Where it meets resistance as it moves on inexorably. The economy will take the battle to society, to destroy if necessary, to conquer if desired. Conquest is at all levels including, fundamentally, the choice of preferred set of values.

Finally, despite appearances, although there is work to do to establish the case effectively, society is not simply reactive. It cannot be since it embraces all manner of human fulfilment, but the way in which it can be proactive, raises as many issues as those prompted by considering the economy on the move. What is necessary now is to consider a number of heavy structures so that we can gauge the strength of the power of change, of resistance and their interplay.

Heavy structures come in all forms. They may be physical, like the existing transport system, they may be a corpus of attitudes towards ways of doing things embraced by, say, doctors, lawyers or teachers. They may also take the form of ideas either in general terms contributing to a better understanding of the world around us or specifically in the field of setting values. Since economic activity is at the heart of our concern a useful starting point is to look at the potency of economic ideas.

Notes

1. Hobsbawm, E.J., *Industry and Empire*, The Pelican Economic History of Britain vol. 3, pp. 58-59
2. Hobsbawm, E.J., ibid. pp. 64-65
3. Hobsbawm, E.J., ibid. pp. 80-83
4. Hobsbawm, E.J., ibid. p. 85
5. Gruson, C. and Ladrière P., op. cit.

Part II
ECONOMICS AND SOCIETY

5 The potency of economic ideas

When Keynes wrote that, 'Practical men who believe themselves to be quite exempt from any intellectual influences, are usually the slaves of some defunct economist' [1], he was perhaps expressing regret that politicians did not know how to make use of contemporary economic thought. He was certainly drawing attention to the power of economic theory. His remark is also a reminder that the exercise of political power often revolves around the ability to exploit economic ideas to political advantage. A contemporary example is the way in which Fisher's theory of money and prices [2] was resurrected in the 1970s and used to underpin libertarian policies. It helped the Conservative party in Britain to capture the intellectual high ground around which most of contemporary British economic and social policy is now being addressed. The political embracing of such economic theories creates a screen of political ideology, which then prevents the theories being tested on their own merits. When politicians call economic theory in aid to support action the theory is often perverted to claim the authority of an unchallenged law; to be obeyed or else.

A body of knowledge, such as that comprised by economics, its hypotheses, its theoretical structure and models, and the ever provisional conclusions emerging from quiet objective study of the field of human experience, does not deserve to be appropriated in this way. Those devoted to the discipline should be allowed to contribute to the pursuit of knowledge in the way that other areas of enquiry, whether they are physical or other social sciences, are allowed to do.

Such, however, is the potency of economic ideas, even wrong ones. So much of everyday life is concerned with economic activity; to produce, to organize or to consume. So much, too, of the political process is concerned with this activity, especially over the use of resources. In the circumstances, therefore, a continuous battle over the ownership of economic ideas is not surprising, nor is the use, and abuse, of those ideas that appear to suit a particular political ideology.

Given this ulterior preoccupation with economic ideas on the part of those exercising or struggling for political power, what ought to be relatively pure research activity becomes tainted by undue regard for whom the ultimate paymasters are likely to be. Economists have always been attracted to the power base of politics. It is possible that the language and presentation of economics itself is biased, by an excessive awareness of the ultimate use of ideas and by cool calculation of self-interest on the part of economists.

An ideological bias may be identified at the relatively trivial level of contemporary political debate, where ill-founded appeals to economic theory are made in order to support immediate political decisions. Such use of economics is an uncomfortable fact of life to be endured. At a deeper level there is a bias such that economic ideas, whether predisposed to free markets or alternatives, contain ideological views locked deep inside them. So deep that they lie undisclosed. Yet they are the values which determine that certain conclusions (apparently derived from theory) appear to be predetermined.

What is meant by ideology? Joan Robinson, in her magnificent short set of essays, says an ideology is like an elephant. It cannot be defined but we know one when we see it. She goes on to say, 'A society cannot exist unless its members have common feelings about which is the proper way of conducting its affairs, and these common feelings are expressed in ideology'. And then she adds, 'Any economic system requires a set of rules, an ideology to justify them, and a conscience in the individual which makes him strive to carry them out'. [3]

There is a continuing danger that economic ideas, because of their ideological roots or because of the proneness of politicians use economic ideas to further ideological argument, play a far more important part than is warranted. In that sense the corpus of knowledge and the attitudes that feed into and off it, represents a heavy structure in society.

Two central, driving forces, underlie economic activity and provide the pillars on which the modern economic model has been built. The first driving force is the individual's and society's concern to improve material conditions. That is to say, to consider what has to be done to ensure that more is produced (and available for consumption) next year than this; this is the crucial issue of economic growth. It is a process that appears to have continued without a great deal of pre-thought or contrivance in western countries over the past two hundred years. It has not happened outside the West (the Pacific Rim countries now excepted) despite the massive amount of thought (and money) that has gone into finding ways of stimulating growth where that has not yet happened naturally.

The second pillar is the quest for efficiency which in many respects is the keystone of economics. Economics is about scarcity. It starts with the fact that at any time the resources capable of being produced by existing factors of production are insufficient to meet existing needs. Scarcity dictates two imperatives; first, that waste, i.e. an inefficient means of production, is to be avoided wherever possible. If one production route makes it possible to produce a television set with less inputs than another, we must prefer the former because the latter wastes resources that could be used to produce other goods and increase the total production of goods. Second, a system is needed by which consumers can exercise choice - to express their preference for lower, rather than higher priced goods; for better, rather than poorer quality goods. The market is the crucial element in achieving an efficient allocation of resources.

Eric Roll [4] provides an admirable illustration of the principal areas of concern for economics.

> It follows that the basic concept of the surplus product or value simply means that human labour is able to wrest from nature more than the bare needs for human survival; that all progress (and civilization itself) depends upon the size of this surplus; and that the division of this surplus between consumption and accumulation and among the various members (or 'classes') of the community is a central economic problem determining, to a major extent, the development of the economy itself.

Roll's summary highlights issues of supreme importance to an understanding of what economics is all about. The first proposition, the ability to create a surplus above what is required for survival (in subsistence economies), or for day to day consumption (in advanced societies), concerns that area of economics devoted to the question of economic development. The size of the surplus is determined largely by technology, normally embodied in capital goods, which is the source of productivity growth (although improvements in productivity can also be achieved by a more efficient use of existing resources). The decision as to the use of the surplus for additional consumption today or for accumulation (to create a greater surplus tomorrow) will determine an economy's future rate of growth. Finally, the task of reconciling the various claims to the surplus raises issues of income distribution; how factors of production are rewarded within the market capitalist model, but also the highly charged issue of who owns the surplus - the stuff of revolutions!

The above formulation of the main issues facing economics pinpoints the features that should be especially monitored in observing an economy on the move. It also underlines the point, especially in considering the competing claims to ownership of the surplus, that economics is concerned with issues upon which society at large will have a view.

The ways of allowing an effective expression of choice in modern economics have been debated long and hard. Systems involving planning, which set social goals to determine priorities and an administrative system to allocate resources, have been practised in a number of countries, but not with a great deal of success except in wartime. In a comprehensive and radical form they have been best suited to countries at the early stages of development, when the bulk of the population is poverty stricken and lacking in basic facilities of health, education and infrastructure. In these circumstances it is not difficult to set social priorities politically that carry the assent of the people. The need for some, any, improvement over-rides fine economic calculation. Thus in the early days of a revolution it can be accepted that everyone, including highly skilled personnel such as doctors, should spend their weekends building roads or contributing to other public works.

At this early stage of development the task is to motivate all to give and to do more for a common cause. But this stage does not last long. The pressing problems of scarcity demand a more refined way of allocating resources. Moreover, as basic needs are satisfied, albeit crudely, the question of determining what people want, the question of choice, becomes more pressing and more difficult to answer effectively without a price system. Even in a modern so-called mixed economy the areas still within the public domain are areas where it is particularly hard to discern what people want; how much health?, how much education? Popular pressure groups and lobbies or the five-yearly (in Britain) general election, fought around vague manifestos, are of little help in setting these priorities.

For these reasons and in view of the intrinsic strength of the approach's internal logic, modern capitalism has embraced the market system comprehensively, both as the best means available by which people can express choice and for allocating resources efficiently. In both these two broad aspects, the setting of a price to clear the market is a crucial tool.

However, each past period has had its own form of economy and supporting caucus of economic theory. At the time the interpretation may have appeared to be absolute, timeless, but in retrospect it can be seen to be essentially relative. The interpretation appropriate to one time becomes irrelevant as circumstances change and provokes a process of adjustment which is often painful and slow. It can be argued that all sciences are relative in this sense. The state of knowledge and the way in which, at any moment in time the scientific explanatory model is formulated, changes over time. In physical sciences the reason for this is mainly a function of the stage of human understanding of the nature of things. That relativism is important in the study of the history of science but it is not, substantially, a product of political or social pressures (although perhaps the church's effort to suppress theories that the world was round produced sycophants continuing to support the flat earth theory amongst the scientific community). However, it would be dangerous for an enquiry into economics to make the same assumption. The current interpretation of economics will, like the physical sciences, depend on and reflect the state of learning at any point in time, but the views held and the way they are formulated will be influenced, additionally and critically, by many other pressures at work in society at large. In some cases the result may be to resist such pressures, in others it may be to run with them too quickly. Whichever way it would be wise to be very much on guard when considering issues of objectivity in economic theory - because the subject is value dependent.

There is another point which relates to the concept of progress over time - an idea precious to our Victorian ancestors and never far below the surface in today's debates. Here, too, the physical sciences are a useful but limited comparison. The succession of theoretical development through, say, Newton and Einstein to modern formulations of physics can be regarded as progress in the sense that no-one would ever expect physics to revert to, say, the Newtonian system, as the preferred explanatory model. In the same way there is clearly much in economic theory that represents progress. An information base hugely greater than, say, that available at the beginning of the century and an enhanced means of analyzing it with computers, have allowed economists to devise and test theoretical systems with practical results well in advance of their predecessors. The present discipline's far more rigorous statements of the conditions for optimal market efficiency, could be regarded as progress, but it is a different matter to judge whether the move, for example, from the compact, socially constrained market conditions of the mercantilist period to the open, allegedly mechanistic operations of today's markets is progress in the same sense of the term. The reason for this is that it is necessary to bring into the theoretical formulations different value perceptions governing the behaviour of both individuals and community. These values are instrumental in the creation and operation of economic models in their respective time periods. When one changes the other will change too. However, whether moving on historically is progress is by no means as clear. There is just as strong an argument in favour of regarding history in this respect as the operation of a dialectic, in which differing perceptions about the balance between individual and community ebb and flow over time and result in different economic formulations.

A proper perspective of economics as a science needs to span at least more than ten centuries, perhaps even more. Indeed, it is significant that Eric Roll, in his definitive study of the history of economic thought, [5] takes the Old Testament and the Greece of Plato and Aristotle as his starting point. He could, perhaps, be criticized for not starting with the more ancient Egyptian civilizations. His scale of perspective is appropriate. Even in Old Testament times there was private property, markets, the use of money and an understanding of the value of the division of labour. Aristotle, too, wrote much about the nature of economics, the principles of exchange and the role of money. Certainly coherent models of economic activity existed in early mediaeval times. These were built on an acknowledgement of the importance of economic activity, both as a process and in terms of the use of end resources. More than now views were based on the human dimension to economic activity identifying the underlying values upon which transactions between people were to be based and were required to be judged. Knowledge was steadily built up which helped interpret cause and effect in the emerging formal economy which utilized prices, a market structure and a wage nexus. Much is to be found in mediaeval literature analyzing in what we would now call economic terms, for example, the consequence for prices, incomes and supply, of a bad harvest. The mediaeval period is also an example of the way in which the circumstances of the time, the dominance of the church over trading matters and its involvement in national and international politics, to a large extent dictated the perspective that was created of how the economy did or should work.

There are even better examples against which to explore the argument. One would be the national and international political scene of the seventeenth century and the accompanying body of thought commonly known as mercantilism. It is an appropriate choice because, in a variety of ways, it represented a clear cut break with the mediaeval period. A second period, even more relevant to the argument, is the period beginning approximately in the mid-eighteenth century lasting through to the middle of the nineteenth century. This period, covering the transition from commercial to industrial capitalism, and the full flowering of the latter, reflected a complete break from both the political and commercial background and the economic models of the seventeenth century. These latter two examples should suffice to illustrate how heavy structures, such as economic ideas, both influence, and react to periods of rapid change.

The reign of James I continued the process of moving out from mediaevalism that had already gathered pace under Elizabeth I. Britain was still a second rate power in comparison with its European neighbours. It had a relatively small population of some four million (at the beginning of the seventeenth century), located largely in the south-east, except for the pockets of population associated with the woollen industry in the west country. Four-fifths of its people lived on the land. The country and the economy was dominated by London. By far the largest city in Britain, London also dominated political and social life. It was also the focal point for the growth in market activity required to meet the city's economic needs.

The pattern of agriculture was largely one of extensive pastoral farming in the west of the country and of arable farming in the east. Farming was mainly for subsistence, but there was a significant export trade. Part of this was eastwards to the Baltic states, but there was also a substantial coastwise trade mainly devoted to bringing provisions and material to the London area. London's population needed to be fed, but the growing craft industries springing up around the periphery of the city, also needed to be supplied and serviced.

Even at the beginning of the century agriculture was changing rapidly. The first wave of the enclosure movement, that began during Elizabeth I's reign, had already damaged the structure of the traditional village community and concentrated the ownership of land in the hands of the upper classes. The movement continued through the seventeenth century, although its purpose was now different, being an attack on smaller independent agricultural workers; to improve arable farming as much as to convert arable land to pasture. The change of emphasis, more consonant with the times of emerging entrepreneurial activity, had less adverse social consequences. It did not, for example, lead to extensive depopulation, and, as it also increased the supply of food, it was more generally acceptable. Nevertheless, the process created radical social change, which largely fed the populist movements of the middle of the century.

Apart from agriculture most economic activity took place in the household. Families made their own clothing or engaged in different craft activities located in the villages or small towns, such as butchers, bakers, cobblers and carpenters. The only industries of any significance were the textile industry and small metalworking firms which were creating pressures for more complex forms of economic organization. The former mostly took the form of a domestic system whereby workers were provided with raw material to work up in their homes. When these activities emerged outside the household, but still operated largely within self-sufficient towns, they lent themselves to the guild system of organization, which tightly controlled the terms and conditions of employment and settled on a mixture of both early trade unionism and business organization.

Externally Britain was barely a trading nation. In 1600 exports and imports together were some £4 million, a small fraction of estimated national income. Almost all of the exports were of wool or woollen goods. The expansion of its overseas trade, and the enlargement of many domestic industries as a consequence, came later. They resulted from the introduction of the Navigation Acts (in 1651 and 1660) which promoted fishing, the transport of trade in British ships and provided the basis for a strong naval force. Benefits to trade and industry were also beginning to flow from imperial expansion (Jamaica was first conquered in 1654).

In international terms, Britain pursued no geo-political strategic overseas policy during the reign of James I. Foreign policy, in the sense of a strategy designed to further Britain's political or economic interests, barely existed. Rather attitudes and consequential actions towards other countries were largely seen as the expression of the crown's own personal likes and dislikes. Thus, regardless of the respective merits of good or bad relationships, James wanted to get on good terms with Spain (because of James' own religious affinities). In contrast he was determined to oppose the Dutch for the opposite reasons. In both cases the national posture was dictated by James' own personal views, with Europe as their focal point.

This situation reflected the contemporary view of the role of the crown in political and economic matters. But it was transformed as the excesses resulting from such whimsical personal rule increasingly conflicted with the interests of other parts of society. The monied classes, who were being asked to pay for the consequences, complained. But there were many radically minded commoners equally dissatisfied with the economic consequences.

An element in the discontent, which was beginning to gain momentum at the early part of the century, was the rapid rise in prices. The rise had begun in the previous century but continued into the seventeenth. It affected all classes. Wage earners, small farmers and those on fixed incomes suffered badly. But the crown and those in Parliament to whom the King looked for additional revenue, were also put into increasing difficulty. These economic factors fuelled the fires of discontent already lit by the consequences of the Elizabethan settlement. As a result a challenge to a royal rule emerged, reflected in the growing rift between king and parliament, in the continuing arguments over the succession and in the increasingly turbulent religious background.

There was another sea change taking place. The end of the Elizabethan period also brought to an end a period of early puritanism which, whilst still resting on mediaeval religious views, did, nevertheless, give a certain dignity to Elizabethan society. The passing of this age resulted from many factors, but not the least was the influence of Shakespeare and his fellow dramatists, who provided a leading stimulus to a more open, more questioning, and, in many respects, a more frivolous society.

Thus the opening of the seventeenth century marked a significant change in British social, political and economic life. The economy remained largely mediaeval, though already being buffeted by the consequences of the first wave of enclosures. There was very little trade with the outside world but a sense of national identity was emerging. More importantly, there were elements in society now discontented with what they had and wanting change. A politically aware popular movement, beginning to sense the power it could wield by combination, was emerging. At the same time the monied classes were seeing less and less advantage in their traditional alliance with the king. It was too early to say that they possessed a positive agenda to direct things in their own way in other directions, but in the negative sense of frustration, acquiescing to regal whimsy was no longer acceptable. The monied classes also saw the opportunities of creating a freer and a more internationally open economy.

Beginning with this destabilized base, the seventeenth century saw a substantial transformation in Britain's economy, which ultimately reflected the changing position and increasing importance of Britain in relation to other European states. The reasons for the transformation are complicated. Some were intrinsically internal, as the British people moulded a new identity and formed new ways of governance - a process that strengthened a sense of national identity. Another reason was geo-political. Thanks to the discoveries of the major voyages of exploration the effective global pattern had been changed. Greater awareness of the East emerged as new sea routes were opened up. The discovery and early settlement of the Americas provided wider world perspectives. As a result of this geo-political shift, the balance of economic power in Europe moved from the countries of the Baltic and Mediterranean peripheries towards those with an Atlantic seaboard. The countries forming the Hanseatic league diminished in strength whilst Holland, France, Spain and Portugal and, of course, Britain, began sharpening their teeth.

The sharper teeth were needed since the process took place with considerable friction that often led to open conflict. Conflict was sparked by protective trade policies, by competition to secure trade routes and by disputes over new possessions or by wars provoked by traditional causes in Europe itself. The primary areas of conflict were in the West Indies, the Canadian and US settlements, West Africa and the Indian sub-continent.

The competition between the British and Dutch was the earliest and most sustained source of conflict. Disputes over fishing rights, rivalry for seaborne trade, particularly around Britain, and the struggle for access to and control over the very rich pickings in the East Indies, created a constant atmosphere of hostility between the two countries throughout the seventeenth century.

Perversely, however, during the Thirty Years War, Britain sided with the Dutch against Spain. Matters were complicated by wars between France and Britain. The complicated origins (mainly religious or of succession) of these disputes can be left aside, but the point needs to be noted that the campaigns included conscious plans for enhancing overseas possessions and, by naval warfare, increasing domination over new trade routes.

The Cromwellian period saw a substantial strengthening of Britain's position in geo-political terms. The Navigation Acts were specifically designed to favour British interests at the expense of the Dutch. Resistance led to an outbreak of hostilities. The consequences were vital for future British interests, for most engagements in the war were at sea. The British retained a fleet built for warfare (rather than, like the Dutch, relying on converted merchantmen) and, after some early setbacks, eventually won the day. The increase in Britain's naval power was now recognised by other states such as Portugal, who were obliged to give British ships right of trading in all Portuguese colonies, and Spain, from whom Britain wrested Jamaica in 1654, and through the Battle of Cadiz in 1657, effectively destroying Spain's ability to wage war for a time.

Further wars with the Dutch in 1665-7 and again in 1672-4 underlined the constant struggle between the two countries to gain sea supremacy. The rivalry with France, growing somewhat later, culminated in the French wars beginning in 1689 and lasting effectively until their successful conclusion under Marlborough in 1710. In this first modern war, European countries fought amongst themselves both on European territory itself and throughout the world wherever they had possessions. The sea engagements were critical. Britain's success led to its huge dominance over colonial territories at the expense of all other European countries.

By the end of the seventeenth century Britain had emerged as a self-assured nation state, strong at home, as a result of internal constitutional developments, and dominant abroad both as a trading nation and as an imperial power.

The transformation of the economy, both in overseas trade and at home, was just as dramatic. The new trading structures, stretching out to Asia and America, led to an increase in the range of products available for consumption at home. Products that were at first the privilege of only a few, were in widespread use even to the point of being considered necessities; tea, coffee, sugar, textiles. The widespread consumption of these products reflected the changing nature of international trade. It was beginning to take its modern dual form of importing raw materials for domestic industry, and exporting final products directly into a mass consumer market. Some important trading items were exceptions and required a different justification. One was the trade in slaves and the other the desire to acquire bullion.

Overseas trade had increased vastly as a result of successful British military and naval campaigns. Imports and exports combined grew by a factor of three during the century. Much of this trade, whilst protected by the flag, was carried out by monopolistic trading companies, of which the Merchant Adventurers, the Hudson Bay Company and the East India Company, are the most well known. Immensely rich and powerful, by the end of the century they had also firmly established a pattern of trading that lavished wealth on the country. The pattern was controlled in both directions; the importation of native products into Britain and the supply of only British manufactured products to the colonies. Moreover, the trade in both directions was carried in British ships.

At home the continued effect of enclosures on agriculture has already been noted. Parliament's success in the constitutional struggle, put more power into the hands of the landowners who favoured enclosures and gave them greater freedom to practise self-interested economic activity. Agriculture had also undergone a quiet technical revolution with more attention to drainage, to the use of manure and better forms of crop rotation. Land law had also changed in favour of the retention rather than the breaking up of land on inheritance. The stability of food prices at the time, in so far as it can be ascertained by data, underlines the achievement of improved agricultural production by the end of the century (although with a reduced workforce).

Putting aside the complicated organizational aspects of overseas commercial trading and their implications for the financial sector, other changes in industry were less dramatic but still significant. By the end of the century Britain still comprised small towns of less than 50,000 people (with the exception of London). Manufacturing was located either on the periphery of the towns or near to supplies of raw materials, such as iron or charcoal, or near to ports to service growing overseas trade. Woollen goods manufacture still held pride of place, but the industry was steadily being transformed by technical improvements affecting dyeing, spinning and warping. Other textile manufacturing was increasing. Silk working came in with French refugees and cotton processing as a result of imperial expansion and access to supplies of raw materials.

Heavy industries such as coal mining, and metal manufacturing, were also emerging as major sectors in domestic production and for exports; so were the traditional metal working areas such as the Midlands and Yorkshire. Significant improvements in transport also took place. Improvements had been introduced in harbours to facilitate coastwise traffic as well as long haul routes. Rivers were made more navigable. Steady progress towards creating a road transport system had begun.

The industrial structure at the end of the century continued to retain elements of the mediaeval system. Craft guilds still thrived and many more were created during the century. In part, the reason for this was the desire of government to raise additional revenue and, in part, the desire of the leaders of craft guilds to secure monopolistic privilege. The continued existence of the guilds suggests that, given the social and economic structure of the time, particularly the dependence on small relatively self-contained markets, their form of organization remained suited to these conditions. However, by the end of the century some guilds had grown to a size where the 'leaders' employed a considerable number of craftsmen, thus anticipating larger and more modern examples of industrial organization and, to some extent, the beginnings of trade unionism.

In the wool and textiles area different forms of organization existed, coming under the generic term of the putting out system, i.e. a system where a capitalist (someone financing the materials required) placed them with workers who, with their own tools and machines, produced the finished goods which the capitalist then bought and sold on. This system had many variations depending on the region (the Yorkshire clothiers were independent producers whilst in the south-west the clothiers employed many wage earners). Heavy industries were, of course, organized more capitalistically, although again with many variations.

The state's view of its role, and what it needed to do to further national economic progress, was reasonably clear cut in terms of overseas trade, but ambiguous and fairly ineffective in terms of domestic activity. Foreign trade was effectively governed by the Navigation Acts of 1651 and 1660. These laws required that trade between Britain and the colonies should be carried only in British ships, i.e. ones built in Britain or in the colonies. Goods could be imported into Britain from other European countries in foreign ships, but would be subject to a surcharge. Through other legislation imports were restricted by levies and exports encouraged by subsidies. What could and could not be produced in British colonies was laid down precisely.

At home the government was disposed to regulation although lacking in strategic aims and resolution to pursue enforcement. In agriculture, landowners were protected by tariffs against harmful falls in prices. To encourage a plentiful supply of wool for domestic production, the export of wool was made a crime punishable by death. Foreign cloth was effectively prohibited by high duties, skilled artisans were forbidden to emigrate whilst foreign skilled workers were encouraged in.

There were many indications of the decline of this closed domestic system by the end of the century. One was the way in which most towns, and above all London, had already developed a substantial trading activity. One consequence was the emergence of a firm money economy which implied the presence of a substantial wage earning class. The markets supporting the towns reflected the existence of a network of trading and manufacturing activity of both domestic products and imported goods. The reasons for the growth of markets at this time are obscure but they probably reflected the growing population, the increase in the money sector as distinct from the subsistence sector, the rise in per capita income, the general acceptance of many goods as necessities for all rather than only for the minority (e.g. tea) and the production of a wider range of better quality domestic goods as a result of technological change.

The seventeenth century acted as a bridge between the mediaeval society of the first Elizabeth and the remarkable sequence of social changes of the eighteenth century. In turn this sequence was the forerunner of the industrial revolution and the creation of modern society. The change of the seventeenth century emanated from both external factors and internal circumstances. In both cases circumstantial change appeared to have set in train a revolution in attitudes, in order to create conditions under which new forces could be allowed to have full rein.

The external circumstances largely related to the geo-political change in emphasis from land-based trade, bounded by the Baltic and the Mediterranean seas, to Atlantic orientated trade, with open frontiers in the newly discovered lands of the Americas and the Far East. In its turn the re-orientation led to the growth of a commercial capitalist system, which reinforced the power of the already established commercial and financial centres of Venice, Amsterdam, Antwerp, and, to a lesser extent, London.

At home the pressures for change were more radical. This was largely because the citadel to be reduced was a more formidable obstacle. It consisted of a set of beliefs still largely dominated by the church where the rules applying to economic activity were no different to the moral standards set for human behaviour generally. This set of beliefs formed the basis for an understanding of inter-personal relationships, of what was fair and reasonable practice, and justified a rigid class and organizational structure. The latter structure was most prevalent in the organization of rural life, built around the parish and the tradition of common land. It was a structure put under threat increasingly by the more modern methods of agriculture and by the enclosure movement. This latter movement did not directly challenge the traditional social organization of a village. However, by creating a surplus workforce, it forced the search for employment off the land and led to a loosening of the economic and social network in the countryside such as was already to be found in the towns. Part of the pressure for change, of which new agricultural technologies were one example, came from developments in scientific thinking generally. Whilst not yet challenging the authority of the church openly, the scientific fraternity was demanding, and getting, a greater freedom of action than was allowed by the mediaeval church's moral teaching.

It would be wrong to exaggerate the extent of the transformation of the internal domestic market by these forces. Economic activity remained not a system, but a mass of individual trades and individual dealings. Nevertheless, the move towards a market economy quickened during the century and new understandings of how the economy worked and, above all, new perceptions of the rationale of the system, were required and offered.

The widening of frontiers following from the New World discoveries provided a dramatic stimulant to the seventeenth century economy. In the first place overseas trading itself created new initiatives that went hand in hand with the exercise of power. Imperial conquest and trading patterns were inextricably intertwined. One result was a substantial increase in military spending, affecting the growth of defence industries at home, such as naval shipbuilding, and the deployment of manpower in the navy and army. Rapid growth in commercial shipbuilding also had a knock-on effect on related industries, such as lumbering and the provision of other shipbuilding materials. The employment opportunities in the newly created colonies were also novel requiring skills in administration and in commercial dealing. Since the latter involved, amongst other activities, funding commercial ventures, which required borrowing capital, charging interest and the allocation of liability in the event of loss at sea, an area of activity was created which had to be justified in ways other than laid down by the mediaeval church's rigid view of usury and related issues.

The growth of venture capitalism opened up the debate about the terms upon which capital could be obtained. The success of commercial capitalism began to create surplus funds for investment which prompted a search for profitable ventures in the domestic economy as well as overseas. In this sense an important lever for prising open the domestic mediaeval economy was delivered by external activity. Also, the substantial increase in wealth of those benefiting from overseas adventures affected the domestic market by creating and demonstrating new consumption patterns and new styles of affluence which many wished to emulate. Taking the domestic and external pressures together, the mediaeval economy was subject to irresistible pressures for change.

At the opening of the seventeenth century the prevailing view of economics was still scholastic, but only just. Writers such as Thomas Wilson were still attacking usury but others were beginning to press for a loosening of ideas and approach. The Frenchman Jean Bodin examined in 1569 the causes of inflation in what can only be called a modern way, listing the supply of gold and silver, the existence of monopolies and the debasement of the coin as the principal causes. More importantly John Hales (1581) adopted an approach which was the forerunner of much to come. He was a humanist and a believer in self-interest, both as a motivator of individual activity and as a guide to state action.

The main attack upon scholasticism developed in a practical way as administrators, politicians and intellectuals all applied their minds to the regulation of trade and industry to further the national interest. The major contributors to economic thinking of the period, Cockayne, Misselden and Sir Josiah Child, were all active political administrators. Much of the theoretical debate was in justification of practical regulations.

The view of economics of the mercantilist period was forged in the context of the growing importance of international trade and the real politik of emerging nationhood. The obvious example was the passing of the Navigation Acts, and related legislation, all primarily designed to enhance Britain's productive power. The purpose was to create the conditions that encouraged the growth of a merchant fleet and optimization of Britain's wealth by seeking to increase its exports and reduce imports. The latter policy was implemented by draconian regulations dictating what type of economic activity was allowed in the colonies and constraining other foreign trade as best it could.

How to manage overseas trade to these ends was less than clear to start with since there was an undue focus on the acquisition of bullion. The desire to hold stocks of precious metals, as an indicator of wealth, was of long historical standing and was supported by direct control over the import and export of precious metals. The narrow view could not stand up against the pressures of international trade and the growth of forms of credit. Much of the theoretical debate revolved around the problem of setting the appropriate exchange rate to ensure the country's surplus in bullion and to curb speculators (where the traditional arguments against usury were brought into play). This theory of international trade was developed extensively by the main mercantilist supporters, such as Malynes, Misselden and Mun - the last two of whom were also active merchants, one a member of the Merchant Adventurers and the other of the East India Company. But it was Mun, in his major work *England's Treasure by Forraign Trade*, who presented the fullest justification of commercial capitalism as seen through mercantilist eyes. Whilst recognizing the traditional view of accumulating wealth, he anticipated a move to the case for maximizing trade - the freeing of which in Britain's later interest, would become the basis for embracing free trade.

In contrast to these external pressures, views about the economics of production remained traditional. Although trade and production had both developed substantially away from their mediaeval forms, the domestic economy was, still, built essentially around the market economy of small relatively isolated towns and consumption patterns largely met by home skills or, through exchange, with neighbours with different skills. In these circumstances the emphasis remained on a tightly controlled set of exchange and production conventions based on traditional concepts of fair wages and just prices. The system continued to be monitored by the craft guild system. It was only as the numbers forced into the cash/wage nexus of employment (by being driven off the land) began to swell, that less personal and more flexible systems of employment and production developed. Nevertheless, the carefully controlled communitarian production system of the mercantilist period, remained in place until it was destroyed by the advent of the factory system.

The mercantilist view, at home and abroad, fitted the facts of Britain's economic and political evolution. Externally the barriers to trade, created by the church-based view of appropriate economic activity, were replaced by a more permissive system, allowing the deployment of commercial capital but nevertheless tightly controlled to further national interest. At home, until the advent of the factory system, the traditional sets of values prevailed and continued to dictate the context for economic activity. The mercantilist desire to control for the sake of national interest, led to control even where that interest was not at risk. The desire for freedom emanating principally from Britain's role in international trade, led to a modification of economic theory which anticipated the release of the flood waters of the eighteenth and nineteenth centuries.

Relevant aspects of the history of the industrial revolution were looked at length in Chapter Two. With a few notable exceptions it was not until the last quarter of the eighteenth century that attempts were made to conceptualize economic processes and to structure such analyses into a coherent view of society. Adam Smith was, of course, the principal British exponent of a comprehensive approach. His contribution was followed three decades later by the equally radical contribution of David Ricardo. Most of what was then written and analyzed on the economy, until the advent of Marx, on the one hand, and the marginalist school, on the other hand, was based on the work of Smith and Ricardo.

The timing of the emergence of this new and comprehensive approach was significant. Britain's role in the embryo world economy was rapidly changing. After a long period in which its power had been enlarged by military conquest, and by the subsequent regulation of trade between home country and colony, the British economy was ready to enter into a second stage of development. A productive capacity, massive relative to other countries, was now in place and it was becoming clear that further gains would follow from freeing, rather than continuing to regulate, trade. The Smithian analysis of the advantages of the division of labour, to exploit comparative advantage, provided the conceptual basis with which to justify freeing international trade. International specialization could be demonstrated to be the best way of optimizing international economic gain no matter that Britain's comparative advantage had been created in the first place by an entirely different route.

Also, by the end of the eighteenth century, the pattern of domestic industry had so developed that a new understanding of the way in which the economy worked was required or, at least, an intellectual rationalization of the way economic activity was being carried out, was needed. The ingredients were all there; processes bringing capital and labour together, a need to understand the process of creating value-added, a need to decide the rules on which the surplus should be divided between the factors of production, a need to understand the operation of the market. One would like to add to that list questions relating to what was required to motivate people to work well, but more about that later. The fact that these processes initially were poorly understood and articulated (particularly in confusing the distinction between value and price), should not detract from the effort that was devoted to providing an intellectual rationale for the new form of economic activity.

As has already been described, the British economy was moving at great speed, from a level of consumption barely above subsistence for many into a relatively affluent mass consumer market. Production processes were now based on new factory-based technologies. Increasingly the mechanism by which such internal trade took place were attracting interest. It was Alfred Marshall who gave a definitive approach to the theory of consumer behaviour (1890), but before that the mechanisms of the market, the determination of price and the equally important need to free markets in order to gain the maximum advantage of specialization, were being recognized.

The combination of these pressures provided a powerful tool for change. It articulated a model of behaviour which legitimized current practices. It created a model based on an increasingly popular libertarian theory. It also gave out a message that businessmen, and all those standing to gain from seizing the opportunities offered by the new system, wanted to hear.

However, already the problems, which were to over-shadow economics and politics in the coming century and later, were there to be addressed. Were the claims of capital and labour essentially competitive? If so, how were their respective claims to a share of the surplus to be accommodated? Was it possible to reconcile them in the relatively safe atmosphere of disinterested economic theory or were the claims the stuff of which revolutions were made? The seeds of these disturbing prospects were sown in the early formulations of the labour theory of value. The marginalist school was able to provide an alternative confident, if not complacent, view of economics. The application of the harsh Ricardian law of increasing misery and its consequences for working class thinking, was followed by the even more systematic and radical thinking of Marx, which painted an alternative and more revolutionary canvas.

The distinction between the two routes which followed the bifurcation in approach, was crucial The marginalist school led to the development of an economic theory seeking to establish objective, positive relationships (in an attempt to mirror the scientific processes of the physical sciences). Such an approach gave scant recognition, if any at all, to either the underlying values presumed by the models or the relationship between the economic model and the wider over-riding social model.

Marx's work would be the examplar of alternative approaches to a wider social model which start with the economic model. However, many sociologists, non-Marxian among them, would take a similar view, seeing economic activity as but one part of a social structure, where the determination of many elements in economics such as wage levels, rates of return on profits etc., are determined or constrained as much by the relative power exerted by classes or other groups as by any identified laws of economics.

The Marxist approach will be put to one side. There is little justification for this great omission, except the urgency and the strength of desire to move on to discuss, in much greater detail, the origins of the prevailing model that emerged from the marginalist school, although more will be said about wider sociological views and their importance to economic ideas later.

The marginalist school [6] provided the means of escaping from the logical prison of the labour theory of value. The approach set the course for the development of economic thought that created the theoretical structure of western market capitalism. In essence the breakthrough consisted of identifying that the key issue to be addressed by economics was that of price and not of value. Given that this is the central issue, interest could then be focused on the market conditions determining the price which would effectively clear the market. Since the analysis needed to explain how a market would move into equilibrium, it was important to explain the response mechanism both from the supply and demand side. The key condition was that market equilibrium would be established when the additional (marginal) revenue from selling an additional unit of product equalled the additional (marginal) cost of supplying an additional unit of product.

Other conditions relating to the smooth working of the market were subsequently added to satisfy equilibrium conditions, many of which now appear to limit the usefulness of the theory. However, once this element of the market economy was put in place it removed the hitherto almost insuperable barriers facing the emerging newly fledged discipline. It was possible to discard the labour theory of value as a means of determining price. The rich distinction provided by Adam Smith between value in use and value in exchange remained, but henceforth the analysis of the factors that determine the 'value' in exchange concentrated on the determination of price.

It also meant that, by creating a more rigid concept of the market, where the supply and demand of products or factors could be balanced, a means for analyzing all economic activity was created. No longer was it necessary or desirable to distinguish between productive labour, for example, that producing goods, and non-productive labour, for example, domestic services. Moreover, it was now possible to treat capital and, later, entrepreneurship, equally rigorously as factors of production. In fact, in the market, as a mechanism for clearing supply and demand, the marginalists were able to offer a tool of wide ranging almost universal application.

The concept of the market, presented by the marginalists, has supported a vast caucus of economic theory, and its policy applications, ever since. It is on this theoretical base that modern economics has claimed to have the means of delivering the major objectives of economic activity; the efficient allocation of resources, low inflation and balance of payments equilibrium, sustainable economic growth. It also provides a rationalization of the distribution of income in which the return to factors of production were determined in a way consistent with other market conditions.

The strength of this approach is revealed when it is compared with those that were subsequently developed and continued to rely on the labour theory of value as the underpinning analytical tool for explaining price and the distribution of income. The differences are, of course, marked. It is common practice to read into the recent collapse of the command economies of Russia and Eastern Europe a total vindication of the western marginalist approach, a conclusion that is justified in the economic sense. The failure of these command economies to build in some form of price mechanism, in order to allocate resources efficiently, has been crucial. In the political sense, the above conclusion is reinforced by the association of the command system with oppressive political regimes.

What is surprising is that a debate between left and right has also been allowed to develop within the market capitalist systems themselves. Given that realistic political choices facing all western societies rest comfortably within the range set by the market capitalist framework, it is curious how highly charged the ideological and political debate has become. The debate provides a useful illustration of the close link between ideologies (and value sets) and preferred formulations of explanatory models. Although the debate is more limited by excluding, for example, a Marxist perception, nevertheless it still reveals the dialectic prevailing in western societies between ideologies; a broadly based libertarian view and a communitarian view. Whilst both schools of thought embrace the market comprehensively their differences are highlighted by the debate over the role of state intervention. The argument over intervention is two-fold. It is about the role of intervention to ensure that the market economy operates effectively, an economic argument. It is also about the role of intervention in order to achieve justice; there is an ethical argument too.

The libertarian view would disapprove of the use of state intervention on both counts. [7] The view is based on a view of natural order (enthusiastically embraced by Adam Smith himself); that society has its own tendency towards justice and order based on a natural sense of justice and self-interest. Also inequality in society, a tempting objective of state intervention, is an inevitable and tolerable result of social freedom and the exercise of personal initiative. Whilst society may need to act to preserve equality before the law, and provide equality of opportunity, it is pointless, indeed damaging to economic efficiency, to seek to impose equality in any other areas. Furthermore, capitalism has clearly worked without much help from the state. Economic growth has raised living standards, has reduced and will eventually eliminate, absolute poverty. No-one should be concerned about relative poverty, since that concern simply reflects envy. Market capitalists also emphasize that the motor of the system is enterprise and the entrepreneur. The risk-taker needs to be applauded and admired and left free to act, both as the essential agent of economic activity and as a setter of trends and taste. Finally, the system revolves primarily around the individual as the decision making unit; not the state or other institutions.

It follows from this set of propositions that those holding such a libertarian position will be hostile to state intervention. They also oppose it because they believe it leads to the politicization of the allocation of resources. The democratic process risks being perverted by encouraging spending to win votes, by listening unduly to pressure groups, by creating bureaucracies and, above all, by increasing dependence on the state.

The above view, which has been particularly sharply expressed during the past twenty years in western societies, and has formed the ideological basis of many governments currently in power, grew out of the awareness that economic performance in the west relative to other regions in the world had been weakening in the 1960s and 1970s. It was allegedly the result of too much state regulation and especially too high a level of welfare provision. The growth in this awareness also followed in the wake of two severe bouts of hyperinflation in the 1970s. As these were the results of major disruptions in oil supplies and not generated internally within the world capitalist system as such, it was somewhat illogically blamed on a malfunctioning of the market. However, hyperinflation knows no equal as a means of frightening middle classes, who are also very adept at searching for scapegoats elsewhere in society.

More in tune with our general theme, it could be argued that the desire to transform the economy was reflecting the build up of pressures created by new technologies, particularly information technology, demanding greater freedom to operate across international markets than allowed at that time by existing national frameworks.

The alternative view, at least one that was in place at the beginning of the 1970s, and which could reasonably be labelled as Keynesian, was also fully committed to the market capitalist system. It accepted the market as the best means of allocating resources efficiently and of providing for the exercise of individual choice. But there were provisos. The alternative view endorsed Keynes' view that uncontrolled capitalism could not guarantee full employment. The labour market would not be cleared in classical fashion (by reducing wages). In any case, Keynes did not regard wages as the determining factor. The key was the desire to invest - a propensity that fell so low at times of depression that no amount of encouragement, either by a fall in real wages or by lowering interest rates, could restore. This theoretical approach, and the strong post-war desire to avoid the recurrence of pre-war levels of chronic unemployment, still argued for a form of contra-cyclical state intervention to maintain full employment. Conventional functions of state to provide a framework of public administration and to satisfy social choice through welfare and education programmes, for example, were accepted as legitimate and desirable activities of the state. There were other grounds for state intervention as well, such as the need for the state to own or control certain industries in order to have the effective means for controlling the economy.

This set of policies could be called the tool of the social democrat. It contained a belief in the virtues of state intervention, primarily to improve the working of the market system, even though confidence in that market system was strong. Keynesian social policy was also concerned to distribute income in ways other than that which would be generated by the market system itself. Part of the case for this policy was on grounds of equity but part was based on the view that the underlying (Keynesian) problem of a lack of sufficient effective demand would be more easily remedied by a more equal distribution of income. The Keynesian alternative view was distinguished as a matter of practical politics and economics, since it sought to demonstrate that the performance of the economy could be improved by state intervention. It was distinguished by its belief that issues of social justice which went hand in hand with a critique of the performance of the economy, could not be left solely to the workings of the market. The fight against poverty, relative as well as absolute, a desire to create a more just distribution of income, a need to impose constraints on market activity and a need to be ready to intervene for the sake of full employment, reflected a different view of how to get the best out of the economy. It also reflected a more direct and obvious link between economics and social values at large.

This brief review of the development of economic ideas has tried to demonstrate two points. The first is that that development has to be seen in the context of larger political and social change. The second is that the historical development has contained a great deal of accommodation. By that is meant that, as the context in which economic activity took place changed, a new rationalization was developed that provided the means to overcome obstacles to change and to justify change. What is extremely uncertain is to decide which were the dominant pressures. Did new ideas provide the motivation for merchants or industrialists to do things in different ways? Or did new circumstances require new theories to aid or justify actions?

Of these two fundamental questions the first has been easier to answer. To compare the understanding of economic activity and the shape of that activity itself at the beginning of the eighteenth century with that of the middle of the twentieth century illustrates the point clearly: when conditions change, theory and rationale follow.

The answer to the second question is more problematical. Accommodation there surely has been over the centuries but whether ideas have accommodated practicalities, or vice versa, is far less clear. Indeed, to get closer to the final answer to this conundrum, the evidence of later chapters enquiring into the parallel development of the ethical framework will be needed. But relativity is clearly a factor and it is important to be constantly on guard against those arguments which seek to impose absolutes - usually as a means of convincing others that there is no choice to what is being imposed. The past tells us different.

An excellent example of the need for an open mind is to be found in the middle of the nineteenth century in Germany. By mid-century Britain was at the peak of its economic power. Despite the contrary route by which it had reached such commanding power, it was now arguing the case of free trade as the basis by which everyone would gain. Friedrich List, the great German economist, saw things differently. In his book, *The National System of Political Economy* published in 1844, [8] he opens by confessing 'to doubts as to the truth as to the prevailing theory of political economy' and his desire to teach others 'by what economical policy the welfare, the culture and the power of Germany might be promoted'.

He then went on to argue that the prevailing theory, especially that deriving from Adam Smith, concentrated on what was good for all nations in concert (compare Adam Smith's title *The Wealth of Nations*) whereas List's concern was for a particular nation, i.e. Germany. His aim was to help Germany 'attain to the same degree of commercial and industrial development to which other nations have attained', which led him to make a distinction between the 'theory of values' and the 'theory of the power of production'. In other words, for a country like Germany entering into the economic game when others had already succeeded, the objective had to be what was good for Germany and to find ways by which a country like Germany could increase its powers of production to match others.

Although List is remembered mainly for his policy recommendations, in particular his infant industry argument, his contribution was far more fundamental because his argument underlines the relativity of economic theory. For a powerful country as Britain at the time, the advantages of international trade were what she sought - in order to exploit the productive power she had already acquired. And the theory of comparative advantage demonstrated that with existing resources, it was best to pursue the gains of international specialization in the way Adam Smith and other free traders argued. For an emerging country the constraint of the limiting factor of existing resources had first to be broken; a tension that continues to exist today. For powerful nations, especially those with a commanding position in, say, information technology, free trade in services is the most advantageous policy to pursue. For weaker countries, rightly wanting to create a productive capacity in these new areas, the problem looks entirely different. [9] Moreover, it can be argued that Japan owes its present dominant position to the application of a Listian view of its development over the past fifty years. It has only recently begun to trust free international trade as a means of generating its future wealth, reflecting the increased degree of confidence it has in its own economic power.

A further example of a way in which an ideology underpinning economic activity can influence human welfare can be drawn from one of the dramatic political economic events of the mid-nineteenth century. In 1847, Ireland, which was then ruled directly from London, suffered the first of a number of years of potato blight virtually destroying its only staple crop. Misery set in and became a matter of immense public debate as to what should or could be done to rectify the situation. The economic facts were unassailable. The domestic crop was lost, the Irish could not feed themselves nor had they any money to buy potatoes from others. Various solutions were proposed, such as subsidizing the price of imported potatoes or other staple foods, but all ran foul of the doctrine of free trade. If that trade were to be distorted, it was argued, the due processes which were required to be set in train (the high prices of potatoes triggering an increase in supply) would be distorted. Eventually, after what has to be presumed to have been honest soul searching, the conclusion was reached in London and annunciated by Charles Trevelyan, the Permanent Secretary to the Treasury, that 'too much has been done for the people'. Ireland must be left to the operation of natural causes'. [10] As a result one million out of a total population of nine million died.

No-one is in a position to judge other generations, even with the benefit of hindsight, as to the way in which they become prisoners of their own ideologies. However, this episode is a pertinent example of how a particular value set becomes enshrined in economic theory and in public policy.

Two major conclusions emerge from the analysis in this chapter. The first of these is that the formulation of economic ideas, that is to say, how men and women have sought to express a view of the way economic activity works, the relationship between cause and effect, has continually changed over time. The way the articulation has changed has also reflected, both in timing and nature, parallel changes in external factors. The second is that, at any given time, the particular set of explanatory ideas has been based on a value set, an ideology, which, far from allowing economic theory to claim a degree of abstraction and objectivity as found in the natural sciences, has grounded contemporary economic thinking in the particular value set of the time. More profoundly, as far as economic activity is concerned, society appears to have had the ability to select from the shelf, so to speak, that particular set of values that justifies most conveniently the courses of action men and women desire to pursue at any one time.

The first point has been continuously reflected through economic history. In the mediaeval period the models of the day were those which described economic activity in the village and the farm, with additional resources being provided as a form of feudal protection either by the state or the church. The principles governing barter prevailed, with great suspicion on middlemen of all kinds. The mercantilist period continued to reflect the smallness, the intimacy, of community work centred around the home. Work revolved around the village or the small town (apart from London). The model was of a single trader or, at most, small groups of workers heavily constrained by rights and duties imposed by guild regulations and by a moral code imposed by the church. Abroad the pattern changed with the growth in awareness of nationhood being honed sharply in war and in struggles over new possessions. With that development came a great stimulus to parts of the domestic industry, such as shipbuilding. It also marked the beginning of commercial capitalism, the emergence of national state regulations and of a need to justify the creation of wealth through overseas activities, in economic terms.

In the nineteenth century, the sheer power of the forces unleashed by the industrial revolution demanded the removal of most restrictions to trade, first abroad and then increasingly at home. By the end of the nineteenth century, a comprehensive economic rationale was in place which has been used to interpret and justify economic activity ever since.

In that sense the twentieth century has been a continuation of the same processes as the previous century. However, there remains a key difference in approach; between the pure doctrine of laissez-faire reflecting a libertarian ideology (an extension of that of the previous century) and an approach based more on community values. This distinction owes much to the strong social and political reactions to the adverse consequences of the industrial revolution of the nineteenth century.

As to the second question, there is much work to do to describe the way in which moral and ethical values largely hitherto, but not entirely, derived from religious sources, find their way into rationales of economic behaviour or, more ominously, emerge out of economic activity. It is clear that it is an area worthy of much more study. Economics is a social science and therefore needs explanations that contain a view of human nature and human behaviour. However, human behaviour, as Joan Robinson reminds us, requires an ideology. Moreover, since economic activity, the production of goods and services, is so close to the process of human fulfilment, of creating happiness either for the individual or for the community at large, it is always intertwined with value judgements. As Joan Robinson again puts it, 'bigger is close to better; equal to equitable; goods sound good; disequilibrium sounds uncomfortable; exploitation, wicked; sub-normal profits, rather sad'. [11]

More fundamentally most basic propositions of economics can be seen to be based on ideology. A pertinent example is the market economist's view of consumer behaviour. It is assumed that choice is expressed by the individual and that the purchase gives him or her the amount of utility reflected in the act of purchase. This assumption reflects a libertarian, atomistic view of society and is in sharp contrast to prevailing sociological views. The latter lay stress on the importance of group action, of group intelligibility and on the value of goods and services in use after purchase.

The development of the discipline of management is another area of considerable interest. As a later chapter will show, a burgeoning problem in economic activity, especially since the advent of the factory, has been how to organize factors of production together, particularly people. However technical the language in which these theories of management are couched, they ultimately reflect views of what motivates people best. Does a threatening or encouraging model get the best out of people? If the latter, how can that co-operative view be reconciled with the over-riding atmosphere of competition in productive enterprise? And what of the motivation to work itself? Is that simply to be seen in terms of financial return or is it part of a wider and intricate complex of personal and social relationships?

What is certain is that this view of the past contains the same awareness of dynamic forces as were identified in examining the way in which new technologies impose themselves on old structures. In this field of economic development, both in practical terms and in terms of theoretical rationales, the same pressures of creative destruction appear to be at work. New circumstances creating new perceptions move in to occupy the ground of old ways and old ideas. The trickle becomes a flood and soon becomes unstoppable. In the process, anything in the old that stands in the way of the new is swept aside until eventually calm is re-established. The reference points then established wait there unsuspectingly until the next tide begins to mount.

Notes

1. Keynes, John Maynard (1936), *The General Theory of Employment, Interest and Money*, Harcourt, Brace & Co., New York, p. 383
2. See Roll, Eric (1978), *A History of Economic Thought*, Faber and Faber Ltd., London, p. 556
3. Robinson, Joan (1964), *Economic Philosophy*, Penguin Books, Harmondsworth, pp. 9 and 18
4. Roll, Eric, op. cit. p. 295
5. Ibid.
6. The school of thought amongst whom Jevons, Menger and Walsas were prominent.
7. I am indebted for the categorization to Bosanquet N. (1983), *After the New Right*, Heinemann, London, pp. 8-24
8. List, Friedrich (1977), *The National System of Political Economy*, Augustus M. Kelley, Fairfield
9. See Seers, Dudley (1983), *The Political Economy of Nationalism*, Oxford University Press, Oxford, for an interesting analysis of development problems seen from the point of view of the developing country.
10. Quoted in Kee, Robert (1981), *Ireland, A History*, Book Club Associates, London, p. 98
11. Robinson, Joan, op. cit. pp. 18-19

Notes

6 Economics and ideologies

'Any economic system requires a set of rules, an ideology to justify them, and a conscience in the individual which makes him strive to carry them out'. [1]

The sets of rules that make up an ideology and the driving force behind their application in western societies, have, for the most part so far, been derived from a religious perspective. Religion is an obvious source for the search for a set of absolutes upon which to base and justify action. However, it is not that simple. As R. H. Tawney has pointed out [2], throughout history, religion (in his example, Christianity) has offered choices upon which to base daily living, as it affects relationships with others and the business of earning one's living.

One choice is to pursue the path of asceticism. The material world is a snare, a temptation. It exposes the believer to temptations to do wrong which he can well do without. The material world also, by occupying his mind and his time on such secular matters, prevents him from devoting himself to that which really matters - the pursuit of spirituality in contemplation.

A second choice is to acknowledge the need to live in the secular world but to take the view that that world is something with which religion has no concern. Believers are in the world, but not of the world. The sin and suffering of the world is only to be expected. The poor are there as they always will be; conscience is not provoked.

A third choice is to accept that religion has something, sometimes, to say about daily living where gross injustice is concerned, but involvement is confined to specific moral causes, the pursuit of which is not intended to involve a commitment to a wholehearted religious critique of personal or community activity.

A fourth choice is to take the comprehensive view of asserting that the whole experience of men and women provides the arena where religion seeks to establish the kingdom of God. All human experience and all human activity thus needs to be brought under the judgement of God.

Even this list may not be exclusive but the existence of such different views does complicate matters. History shows that differing emphases are placed on these options over time or, to confuse matters more, the views appear to be able to co-exist at the same time. What is a religious view at any one time is thus usually extremely difficult to pin down. This difficulty is in stark contrast to the inclination of those concerned with religion to talk in terms of absolutes and to the proneness of believers to have an unquestioning confidence in the rightness of their grounds for action or inaction.

The mixture of co-existing, but conflicting, stances, accompanied by assertions of absolute rightness, has to be placed alongside the historical evidence which suggests that humans have a relatively easy faculty for changing the emphasis, if not the beliefs themselves, when external circumstances require. The noble thought that beliefs develop and ways of life follow is not easy to demonstrate. The alternative is that men and women, as individuals and in concert in society, have an uncanny knack of making the ideology to fit the convenience of the time.

There is thus nothing to suggest that the task of examining the roles of values and ideologies in economic and social development will be either easy or comfortable. It is a complex area obscured by the passion that often accompanies the defence of ideologies. Moreover, history presents itself as a long continuum of change. To make sense of it, it is necessary, therefore, to select, hopefully wisely, periods around which to develop the argument which can at least be justified by the fact that they appear to be key turning points.

As it happens there are three examples that conveniently are parallel to the historical periods upon which attention was focused in earlier chapters. The first is the period of the seventeenth century; the religious and political ferment surrounding the civil war which pulled society out of the mediaeval period was parallel in time to the mercantilist period. The second is when capitalism was in full spate at the end of the eighteenth century and the beginning of the nineteenth century. The third is the modern debate which needs to be conducted in the context of the changes accompanying the new service economy.

The trigger point of the first half of the seventeenth century was the struggle between king and parliament, which developed into conflict between the Royalists and the Parliamentarians, ultimately forcing the two sides into civil war. The prime distinction between the two parties revolved around respective views of where political power should lie. However, the division also reflected economic and other changes taking place in society, particularly the rise of a merchanting and farming class that challenged the aristocratic leadership of the day. The climax came in the ultimate defeat of the Royalist forces, the beheading of Charles I and the introduction of a period of commonwealth government of immense significance to the future development of British society.

The period was characterized by a revolutionary ferment unique in British history. It focused on the ideological and political positions taken up within the Parliamentarians' New Model Army, an agitation exacerbated by the disenchantment over lack of pay, etc., which followed hard on victory in arms. The pressure for a more democratic and egalitarian society is fully revealed in the well-known debates of the army council. But the consequential action went beyond the army, especially as soldiers returned to civilian life, often poverty stricken, owed pay and retaining the new political views. Splinter movements, such as the egalitarian Levellers and the land sharing Diggers, both reflected the debates and inspired the actions of common people. They represented a distinctly stronger involvement of common people and can be seen as the beginning, perhaps, of working class ideology.

The impetus of these ideas was restrained by Cromwell himself, who was no revolutionary, and by the Restoration, which reflected a view that the Puritan culture was too oppressive. However, future generations inherited an irreversibly new view of society that owed much to this revolutionary period.

These political developments were inextricably interwoven with the religious controversies of the period including Henry VIII's break with Rome and the consolidation of that break under Elizabeth, after the attempted return to Rome by Queen Mary. Attempts to regularize matters were made in the Act of Settlement of 1559 and in the following Act of Uniformity, which reinstated the Prayer Book and required an oath of allegiance from clergy. Although there remained a smallish Roman Catholic population and a few fanatical non-conformist sects, such as the Anabaptists, this political compromise was designed to accommodate three mainstream views of religion and church organization.

One party saw in the break with Rome a necessary act to restore religion to its true state. It held the view that the Church of Rome had lost its way and that the Church of England, thanks to Henry VIII, could and should be regarded the church which maintained true links with the early church of the fourth and fifth centuries. The Church of England was thus claimed as the authentic apostolic church; broadly speaking this remains the view of the Anglo-Catholic wing of the Church of England today.

The second party can loosely be called the Puritans; people who wished that the Elizabethan purge of the Roman Catholic church could have gone even further than it did. They wanted the new church to be entirely free of the Roman tradition; a church purified (hence the term Puritan). For this group the Act of Settlement of 1559 did not go far enough as it allowed the Church of England to retain many of the doctrines and practices of 'Rome', which emotionally became to be seen as the 'Anti-Christ'. They argued, in effect, that there was no logical stopping place between, figuratively speaking, Rome and Geneva. The unfinished task was to move from the Roman communion to the Geneva fortress.

The debate was about all aspects of church life; priestly dress, church ritual, orders of worship and theology, but it also bore on church and state relationships. The settlement sought to restore a church and state link; an issue on which the Puritans, in a sense, held an ambiguous view. They were opposed to the specific arrangements under the settlement but they were less certain of what they wanted in its place. The ambiguity, to a large extent, reflected a difference between the views of the two inspirers of the new nonconformity, Luther and Calvin. The Lutheran view rejected, not only the domination of the church as it was practised by Rome, but also wanted to resist any form of ecclesiastical domination. Their view formed the basis for churches opposed to state control and were inclined to disassociate themselves from involvement in society at large. On the other hand, the Calvinistic tradition, dramatically practised in Geneva, had as strong a view of church and state as Rome or the Church of England, even if the episcopal tradition was replaced by governance at local level by elders (offering to believers, who managed to travel the long hard road to salvation, all the privileges of the elect).

Alongside this alternative but ambiguous view of church and state the Puritan movement also attracted an increasing number of people who protested strongly against greed and riches, both in the church and in society at large (reflecting the egalitarian pressures of the commonwealth). This ostentation provided a further reason for opposing the established church - a distaste for the riches flaunted by bishops and other senior clergy.

Finally, there was a party that can best be described as sectarian, or independent, who sought to abandon the rigidity and form of the Roman/Anglican tradition. However, it was reluctant to replace it with the equally rigid structure of the Puritans. Its position was both theological and ecclesiastical, but its proponents were also part of the rebellion against an imprisoning mediaeval puritanism (a reaction encouraged by Shakespeare and other dramatists and consummated in the Restoration). There were many components of this group, the Latitudinarians, the Quakers, the Ranters, the Antinomians, who all reflected a more relaxed nonconformist view than that expressed by the Puritans and were much more in tune with their time.

By 1640 the first group was totally linked with the Royalist cause whilst the other groups sided with the Parliamentary cause. As a consequence of the victory of the Parliamentary forces, therefore, Puritan beliefs and practice were imposed generally, for example the abolition of bishops, strict Sunday observance and restraints on most forms of pleasure. There was also a move towards a Presbyterian form of church government, partly to satisfy the Scots whose army had been needed to defeat the Royalists. The Westminster Assembly and the Solemn League and Covenant (opposing, amongst other things, the Book of Common Prayer) led the whole of Britain to the verge of adopting a Presbyterian form of government. London, for a while, did adopt the system.

Many were not attracted to the idea of replacing the autocracy of the Pope by that of the elders. Opposition arose particularly within the New Model Army, where new found democratic ideas fitted much more closely with the sectarian/independent religious view. The sense of liberation grew in strength in intellectual circles, in the army and in parliament and the church. When Milton wrote, 'Give me the liberty to know, to utter, and to argue freely according to conscience, above all liberties' [3] he was speaking for many in society. Such was the strength of this movement that the sectarians (or independents) grew from a handful in 1640 to control the army, force the execution of the king and found the commonwealth.

It was clear by the middle of the century that important class distinctions were allying themselves, on the one hand, to the sectarians and, on the other hand, to the Presbyterians. The sectarians found amongst their ranks what would now be called lower middle class people, e.g. mechanics and artisans, who worshipped without dignity or restraint, frequently interrupting services and claiming revelation. By contrast, the Presbyterians were mainly merchants and monied classes enjoying the tightly controlled church discipline with an aversion to 'enthusiasm'. The distinction was strengthening and was creating a link between belief and political agitation. The Levellers and the Diggers saw little distinction in origin between their desire for religious freedom and their desire for radical democracy. 'I am sure that there was no man born marked of God above another.' [4] 'We have a free right to the land of England, being born therein.' [5] The Presbyterian wing observed these developments with alarm.

The ferment over religion continued unabated. The increased toleration of the restoration was followed by the application of the Clarendon Code, created over the period 1661-1689, and revolving principally around the 1662 Act of Uniformity, which made dissent lawful, but disadvantaged the dissenters. Dissenters were excluded from power and from access to, for example, the universities of Oxford and Cambridge. The Presbyterians were particularly hard hit as they aspired to higher things socially. In England, they drifted into Unitarianism - 'a feather bed to catch a falling Christian?' [6] Those priests who would not accept the Act (1,760 vacated their living) provided the main impetus to the formation of 'nonconformist' churches. Full recognition for nonconformists did not occur until 1871, when the same rights were also given to Roman Catholics.

Puritan theology had an enormous impact on British religious, political and social thinking in the seventeenth century. Its Calvinistic approach has formed the basis of individualistic/libertarian action ever since. The puritans identified two covenants. The first was the contract between God and Adam by which Adam vowed to keep the law. In default, he, and all humanity that followed, was plagued by original sin. The second covenant which God offered was of a different sort because it was a unilateral offering by God of his Son to redeem the elect - only a few would be chosen to be saved (c.f. the parable of Dives and Lazarus). The elect were saved not because they believed, but believed because they were saved. Once saved, always saved, claimed the Puritan (although a contrasting view, held by the Armenians, allowed for the possibility of backsliding).

The principle of election did not make matters any easier for the Puritan because his spirit had to be awakened to the fact that he was saved; it was the task of preaching and intense soul-searching by the individual to convince himself that he was saved. It was not an easily achieved task. It was not by works but by determination. 'Believing is sweating work', said John Bunyan, 'the sinner had to run for heaven, fight for heaven, labour for heaven, wrestle for heaven'. [7]

Once the Puritan was convinced he had gained the prize, his position was unassailable. He could reflect on the virtues of his own labour and the pilgrimage by which he discovered that he was saved. The continuation of that effort would be a constant reminder to him and others of his elect position. He would judge others by their results too. Society had entered the era of the individual - God-fearing, hard working and demanding freedom.

The second example of a period in which there was a major interaction between religious ideas and economic activity, was the industrial revolution from the mid-eighteenth century onwards. R.H. Tawney's masterpiece *Religion and the Rise of Capitalism* provides a definitive analysis of the relationship between religious ideas and the emerging structure of economic activity. Reference has already been made to the choices open to believers; the detachment of the ascetic, the disdain for the world, of the other-worldly, the half-hearted acknowledgement of social injustice and the total embracing of worldly affairs within God's kingdom.

The early mediaeval scholastic view, whereby economic activity, like all human activity, fell into a natural order under the command of God, was an example of this last category of embracing all worldly affairs within God's kingdom. There was a divine plan which embraced everything. 'Grace works on unregenerate nature of man not to destroy it but to transform it.' [8] The over-arching plan demanded obedience both from the individual and from society at large.

The mediaeval religious position recognised the inevitability of privilege, of inequality, of a certain harshness to life. This view of the nature of the social nexus and class structure, was rationalized by the theory of the human body. Each component in the body had its place, its function. 'The rich man in his castle, the poor man at the gate' as the old hymn put it. There was an inequality between classes, even though it co-existed with equality within classes. It was a harsh, but nonetheless protective, system since the class structure carried responsibilities as well as rights.

Tawney also makes the point that the earlier system also provided a 'doctrine of economic ethics'. [9] Economic activity (filling an enormous part of daily life) could not be left to the devil. It had to be brought under the sovereignty of God. Although far from the complexity of the later industrial capitalist structure, the earlier society still comprised a mass of individual traders and dealings responding particularly to the growth of city life. To make sense of this mass of activity a number of basic rules were constituted that governed economic life for well over a century. According to Tawney, the basic premises were twofold, 'that economic interests are subordinate to the real business of life, which is salvation, and that economic conduct is one aspect of personal conduct upon which, as on other parts of it, the rules of morality are binding'. [10] These two principles imposed severe limitations on practical activity. In the case of wealth creation, 'It was right for a man to seek such wealth as is necessary for a livelihood in his station. To seek more is not enterprise but avarice and avarice is a deadly sin'. [11]

As regards the business of engaging in trade, the activity was acceptable but, 'It is a dangerous business. A man must be sure that he carries it on for the public benefit, and that the profits which he takes are not more than the wages of his labour'. [12]

Ownership was a regrettable necessity. 'It is to be tolerated as a concession to human frailty, not applauded as desirable in itself'. [13]

The overall message was that all should be aware of the risk of committing the sin of avarice. To avoid it, it was necessary to comply with the ethical basis of economic transactions. Prices needed to be just (be warned by the fate of Shylock!) and wages needed to be fair. Hence the growth of regulations controlling prices, the contemporary view of usury and, through the tight monopolistic guild system, the regulation of wages and conditions of work.

The need for an ethical basis of economic transactions revolved around the view that there were great moral dangers surrounding economic activity particularly if moneymaking became obsessive.

How were those holding this set of ideas and values existing at the beginning of the seventeenth century (this heavy structure) to respond to the radical changes in economic activity occurring in the second half of the eighteenth century? One response would be to denounce the impact on the existing ethical structure as a relapse into pagan immorality. Alternatively, the sea change could be harnessed to provide a reinterpretation of the means and route to be pursued by people towards their spiritual end.

The Lutherans [14] mainly took the first position; to turn their backs on the changes and to seek refuge in a form of individualism. Perspectives and horizons were radically cut back and emphasis put on simple things; honest toil, family life and personal salvation. For similar reasons, the tradition accepted the existing social hierarchy and held the state, not the church, responsible for social order. Faced with the theological choice of whether to love God first and then men, or to love God through loving men, Lutherans opted for the former. Good works were not necessary. Secular interests had little religious significance.

The Calvinists [14], argued Tawney, although seeming to start from the same position as the Lutherans, took a radically different approach. 'Where Lutheranism had been socially conservative, deferential to established political authorities, the exponent of a personal, almost quietistic, piety, Calvinism was an active and radical force.' [15]

Calvinists began, therefore, to see things through the eyes of the practical businessman, adapting to the new economy and for the most part doing very well out of it. As a result the earlier premises upon which economic ethics had been based were crucially revised.

First, Calvinism no longer accepted any ethical discrimination between profits and wages. It recognised the need to avoid the greedy exploitation of opportunities created by risk, but saw nothing wrong in what had previously been called usury (that bête noire of scholastic tradition) which was stripped of its adverse ethical implications. Like the Lutherans, Calvinists were strongly in favour of men disciplining their characters by hard labour which they construed to be a devotion to a service acceptable to God. Business and economic activity was seen as a major context for salvation. 'That aim is not personal salvation, but the glorification of God, to be sought, not by prayer only, but by action - the sanctification of the world by strife and labour.' [16] 'Good works are not a way of attaining salvation, but they are indispensable as a proof that salvation has been attained.' [17]

Thus Calvinism stood 'for a new scale of moral values and a new ideal of social conduct' [18] that served to liberate the new middle class business community. It produced an 'insistence on personal responsibility, discipline and asceticism' [19] and a call to fashion social institutions to match. It is not true, however, that the Calvinist view at all times supported an individualistic view. The early experiments in Calvinistic theocracy in Geneva were based on communitarian not individualistic perceptions (compare, too, the contrasting swing in the utilitarian movement in Britain from a libertarian base to become the foundation for welfare economics). Other nonconformist denominations, not based so rigidly on the continental reformers, such as the Independents and the Methodists, also provided a bulwark against extreme individualism. Nevertheless, the prevailing ideology, and the practice that went with it, of this new and growing middle class, created what have come to be regarded as Victorian values.

This new value base underpinned the perception of the winners in the system. What of the losers? What attitude, for example, did the winners adopt towards their less fortunate brothers? Unfortunately it was a harsh one. Calvin quoted St Paul 'If any would not work, neither should he eat'. [20] Conduct and action were proof that the gift of salvation had been accorded. The poor were, in these terms, clearly failing. The reason for this failure lay in their own attitude, their own lack of resolution, their own weak determination to carry out the necessary struggle to determine whether they had been given the gift of salvation. The poverty of those who fell by the way was not misfortune but moral failure. Riches, on the other hand, were a blessing, a reward for energy and will.

This ethically based position, powerful as it was, contained, as Tawney pointed out, two key paradoxes. First, 'To urge that the Christian life must be lived in a zealous discharge of private duties - how necessary! Yet how readily perverted to the suggestion that there are no vital social obligations beyond and above them!'. [21] In other words, it was one thing to argue that a key element in the Christian life was the practise of private duties, but it was another to go on to refuse to accept that faith led to social obligations. Second, 'To insist that the individual is responsible, that no man can save his brother, that the essence of religion is the contact of the soul with its Maker - how true and indispensable! But how easy to slip from that truth into the suggestion that society is without responsibility, that no man can help his brother'. [22]

These new ideologies underpinned economic activity into the nineteenth century, and were parallelled by other contemporary developments in ideas. A notable one was the great debate on evolution, centred on Darwin's theories, and the rise of a related sociological view (as presented particularly by Herbert Spencer). What these streams of thought had in common was a common view of an evolutionary process, which brooked no weakness in human character. The term 'survival of the fittest' was attributed to Herbert Spencer and referred to the argument that there was a natural selection process within the human species as well as between animal and plant species. 'People who multiply beyond their means take 'the high road to extinction'; they die off in droves, as 'we have already seen exemplified in Ireland'. Those who remained are 'the select of their generation'. Having exercised moral restraint and foresight, they bequeath their powers of 'self-preservation'. [23] In Spencer's terms successful people (in business) were superior to unsuccessful people by virtue of their success. Thus they deserved to survive and were a part of a process whereby the stock of society will be pruned of its weaker and inefficient elements. The closeness of this to Darwin's theory of evolution was obvious. What was less obvious, although the point is made in Desmond and Moore's recent biography of Darwin, was that Darwin was attracted to the laissez-faire argument of Adam Smith as a model which revealed how the weak were sifted out from the strong. [24]

The argument of this chapter so far, and the message to be drawn from the developments of the two periods to which particular attention has been directed, is that economic development and changing values are closely linked. Leaving aside the issue of the direction of cause and effect, it appears that during periods of rapid economic development issues of ideology emerge. They are to be seen in the search for a new ethical basis to justify the actions of those who run with, and take advantage of, change. They are, also, to be found participating in the process of destroying old or creating new heavy structures, constraining, or enabling, new forms of economic activity.

The conflict that arises is as relevant and material to the contemporary surge in economic transformation which has been described in detail in Chapter Three. Modern capitalistic society, whilst grounded in the past, is now dominated by economics and by the power that the command over economic resources provides. The bulk of society's ends are now materialistic and the means to achieve them are delivered by a highly complicated technological system. The system dictates a framework within which men and women have to operate if they are to reap the rewards from the system. It is a society with strong power structures that, either through conflict or through struggles for empire, have played a crucial role in forming and forging much of society. In Europe it has especially led to the emergence of a dialectic between the left and the right, to use an ideological shorthand, which has formed the basis of the political framework of most European countries. The choice of the value basis upon which to establish goals and govern the means of achieving them is just as crucial today as ever.

As far as Britain is concerned the form of the debate between left and right has been peculiar to this country. There are many reasons for this, some of which have already been rehearsed. A relatively open society emerged very early on, as did elements of a democratic process. There was an absence of feudalism and serfdom in contrast to most of continental Europe. Above all, Britain was the first European country in the field to embrace at the end of the eighteenth century not only the economic consequences, but also the wider social changes required by the industrial revolution.

It is a peculiarity of British development that a Marxist solution has never been seriously proposed, even though much of the data used by Marx related to Britain. While Britain experienced exactly the conditions for the revolutionary change expected by Marx and his followers, it has never occurred. Many reasons have been put forward for this puzzling fact. A detailed answer lies outside the scope of this study. Some have argued that the religious revivals of the nineteenth century deflected revolutionary views. Perhaps the longstanding sense of democracy and the relative efficacy of the rule of law (with exceptions during some periods of intense repression), prevented a revolutionary head of steam being created. Also the traditional outright opposition between the landowning class and an emerging proletariat experienced elsewhere, did not exist in Britain.

However, the strength and power of radical movements must not be underestimated. There is a rich seam of radical thought and action which can be traced through from the great mediaeval cry of the Peasants' Revolt 'when Adam delved and Eve span who was then the gentleman?', to the turbulent days of John Bunyan and the commonwealth period, to the emergence of populist movements of the eighteenth century and to the working class movements of the nineteenth century.

Given the success of these movements, particularly the emergence of the trade unions and their ability to convert their beliefs into a political movement, which ultimately formed the basis of one of the two main parties in the British political system, few may regret the absence of more radical solutions. The current weakness of the ideological basis of Britain's traditional left, however, given the political dialectic of its history, is a matter of supreme importance that needs to be addressed. The contemporary articulation of the view of the right can be relatively simply stated but it is much harder (although just as necessary) to identify a contemporary expression of the left.

That expression is rooted in British history although the modern view owes most to the developments of the nineteenth century in response to the social conditions created by the industrial revolution. One tradition deserved the title of utopian. It consisted of a series of experiments designed to reflect and practise ideal concepts of working communities. Outstanding examples in the nineteenth century were Robert Owen's venture in New Lanark, the community of Saltaire and the creation of Port Sunlight. More hard headed than these experiments were the sustained campaigns of the Chartists, the growth of unions, the spread of adult education and eventually the emergence of the Labour party.

These movements, most of which were in one way or another created in reaction to the gross economic and social distress created by industrialization, developed a critique of capitalism and provided an ideology for alternative economic models. The process was essentially political but it was accompanied by a parallel development of religious thinking which provided an additional radical justification for its ideology. The latter development can be loosely encompassed by the term Christian Socialism. [25]

The reason for the emergence of a Christian Socialist movement [26] in Britain differed in one particular respect from the reasons for the birth of similar movements on the continent. There, a real threat to society was perceived to lie in the growth of communism and the Christian Socialist parties were largely prompted by the Roman Catholic church to counter the threat. In Britain the Christian Socialist movement saw its enemy as capitalism not communism.

The Roman Catholic church in Britain became heavily involved in social issues in the middle of the nineteenth century, as the numbers of Irish migrant workers grew rapidly and who worked in extremely harsh and unsafe conditions or were subjected to extreme poverty when unemployed. The church, particularly under the leadership of Cardinal Manning, quickly aligned itself with the plight of the poor. There was also much activity at the other extreme amongst the evangelical wing of the Anglican church. Lord Shaftesbury was known as the 'poor man's friend' although the evangelicals on the whole, approached problems of poverty in the context of the Calvinistic view described earlier. They were an excellent example of Tawney's third religious category - those who were not inclined to accept that religious belief demanded a comprehensive commitment to secular affairs but were nevertheless driven by conscience to react to gross injustices.

The main impetus within the Anglican church came from the Anglo-Catholic wing. Drawn into the East End of London to minister to those living in such awful conditions, the priests and laymen involved organized themselves quickly. One of the first groups was the Guild of St Matthew founded in Bethnal Green in 1877 by S. D. Headlam. The group became the Church Socialist League and later the Church Socialist Union. Under the leadership of Bishops Westcott, Goss and Scott-Holland, it eventually became the Christian Socialist movement which continues to exist to this day. The Anglo-Catholic development was matched by equal activity amongst the nonconformist denominations reinforced by the creation of city missions of the Methodists and the Baptists and those that split from the Methodists, impatient of the seeming lack of concern for the poor, to form the Salvation Army. Gradually these semi-political church movements became ecumenical. A further well documented development was close links that existed between non-conformists and local trade union leadership. The organization of many trade unions at local level was said to be based on the Methodist class system. Certainly many a trade union orator learnt his trade as a local preacher in a nonconformist church.

By the turn of the century and persisting through to the 1960s, there existed an articulated value base for the left position of British politics. Of course, many contributed to the creation, maintenance and development of this position, but one of the most notable contributors was Tawney. His academic work *Religion and the Rise of Capitalism* has already been extensively referred to, but his active life in the adult education movement, his further writings, such as *The Acquisitive Society* [27], and his key role in Christian and political thinking on the left, gives him an unrivalled place in recent British history.

Tawney's life and times have been sensitively summarized by Dennis and Halsey. [28] Tawney was born in 1888 and died in 1962; his active adult life straddled the whole of the first half of the twentieth century. Born in

Calcutta, of upper middle class parents, he went to school at Rugby and thence to Balliol. He was intensely religious and thus, as Halsey puts it, from an early age willingly accepted the social obligations that were the corollary of privilege. He quickly aligned himself with kindred spirits such as Ruskin and went to live for a number of years in the East End of London in Toynbee Hall. His career was moulded by a long spell with the Workers Educational Association interspersed with spells as an academic, first in Oxford and then more intensely at the London School of Economics.

His initial excursion into the adult education field, when he went to work as a regional officer in Rochdale, gave him his first encounter with the common man and his ideals. But this experience was forcibly reinforced by his experience in the First World War, where he served as a sergeant in the Manchester Regiment. He was wounded and invalided out but the experience left indelible impressions. He developed a strong conviction of the evil nature of the system that created the conditions for war. He also formed his admiration for the common sense character of the British working man alongside whom he served in battle.

It was out of these experiences that Tawney developed his own views of economics and society. The basis of his ideological position, which derived from his own Christian conviction, was simple. A moral transformation of the individual, i.e. conversion, was a prior condition. If it were achieved it would then be possible to liberate and use the many talents of working class men particularly in industrial life. Such 'converted' men and women would also bring a more responsible approach to the use of resources and especially consumer patterns. Conversion and adult education combined would thus lead to a good, virtuous and responsible style of living.

Tawney regarded capitalism as anti-Christian. He believed that under capitalism economic means had been converted into over-riding ends. Capitalist society had thus created false gods. He favoured a socialist view because only socialism could put the quality of life above the quantity of possessions. He shared Ruskin's view that 'the only wealth is life'. Whilst Tawney's religion was intense and personal, as he expected everyone else's to be, he believed that society as a whole had a key role in creating the conditions within which individuals could flourish.

On a personal level Tawney's lifestyle verged on the austere (very like Stafford Cripps and other Christian Socialists in the same mould who followed him). Halsey tells a story of the time when Tawney invited Archbishop William Temple to supper; at the appropriate moment Tawney pulled a plate of two cold pork chops from behind some books for them to share. [29] Tawney became angry when people talked about what a person was worth, 'What a person is worth is a matter between his own soul and God', he declared. [30]

Although standing head and shoulders over most, Tawney was still in the mainstream of left wing reformers. In the British tradition he was, as Halsey points out, a steadfast political realist. He played an important role in saving the British left wing tradition from the determinism of Marx. He also stood apart from the views of many later sociologists, by stressing the power to change that came from will and character; particularly if, as he wished and hoped for, a person was motivated by Christian belief. [31]

The debt that modern politicians of the left owes to Tawney is incalculable. Much of the credit for creating the value basis underlying the social democracy practised after the Second World War, must be given to him.

As important an element in the broad view of values of the nineteenth century, was what might be called the theology of the winners. By the middle of the century, there was no doubt about the sense of pride and achievement enjoyed by the growing number of people who were benefiting from participating in economic activity. There was a bold aggressive approach to life and to hard work, often resulting in an austere view that appeared to enjoy the task of working more than the fruits of work themselves. Work was seen as a service to God and that service did not require much regard for the plight of others.

As the century wore on, that positive and overly triumphalist view steadily weakened. There was a greater awareness of the cost of economic progress. There was an increasing sense of the competition being created as other countries struggled to find a place alongside the (up to then) dominant world nation. The cost of maintaining Britain's prime international place in the world increased and the fanfares of praise for the Victorian vision became more and more muted. In the twentieth century the terrible wars, together with deep depressions, combined to sap the confidence that had previously reigned supreme.

The emerging Labour party, as well as the many other groups of opposition, naturally took a more sceptical view of the balance between costs and benefits of the new market capitalist model. The Liberal party had moved away from its free trade position, as far as external trade was concerned, and at home took an increasingly interventionist position to deal with social problems. The Conservative party never really freed itself from its long tradition of paternalism. The continued domination of its policies, by landowning and other non-industrial sections, led to it retaining its patronizing view towards industry, holding 'trade' in relatively low esteem. Wiener [32] dramatically identifies the flaw in British cultural attitudes to business activity shared by all parties (and especially the Tory party) during this period.

Against this general cultural background, of antagonism from the left and indifference for the political right, those in business could at best expect to be allowed to go about their affairs in a kind of value neutral mode; a 'business is business' approach so often found in the United States. This reflected an acceptance, but not an endorsement, by the community at large that it was important to allow businessmen to run their businesses as they saw fit, since the fruits of industry, wealth creation, were wanted by all. However, no-one was required to try to elevate such activity onto a high moral plane, other than to praise hard work and realism, and to accept the application of 'vices', such as greed and avarice, as if they were virtues.

The resulting conditional acceptance of market capitalism was dramatically challenged in Britain in the 1970s. Since then there has been a period of intense rethinking of the underpinning ideology for market capitalism that still remains firmly in place. Although it is relatively easy to track the course of events since the 1970s, as the debate has developed, it is not so easy to explain why there was such a sudden break with tradition. As was referred to in the previous chapter, a number of candidates offer themselves as explanations for the shift in view. One relates to the literally frightening experiences of the two periods of hyperinflation in 1973/4 and 1975/6. Whatever the true causes of the inflationary conditions both periods did enormous damage to the social fabric. They adversely affected incomes for those on fixed incomes or in professions without a militant approach to wage bargaining. Even where wage bargaining was well established, since strong unions fought hardest to preserve real incomes, the traditional consensus view and sense of justice in the practice of income distribution broke down comprehensively. Also, these events coincided with a growing concern that many of the existing institutions, created by welfare state legislation, had become bureaucratized and were being controlled increasingly in the interest of those supplying the services rather than in the interests of those requiring care. Moreover, those supplying services were seen to be under the control of strong unions. The flouting of union power, as a result of the scramble to avoid suffering from inflation, was thus reinforced by the irritation shared by many of being confronted by 'unionized gatekeepers' when the need to approach the alleged care institutions arose.

More importantly, the pressures being generated by the new wave of technologies were also reaching bursting point by this period. It was not only the pressures created by information technology but also the switch away from many (often highly unionized) traditional industries, for example, from coal to oil, from rail to road and, in the face of international competition, the decline of steel making, ship building and other heavy industries. A desire for greater freedom in order to exploit new possibilities was in the air.

Interestingly this period of radical change focused very quickly on a debate over ideology. One aspect of the period of preparation that the new Conservative party undertook in opposition during the 1970s, was to draw heavily on the work of the right wing intellectual schools in the United States. The writings and debates of the Chicago School of Economics, and especially the contribution of Professor Milton Friedman, were closely studied. Working groups, and subsequently small research institutes, were established not only to work through the ways of applying the policies already in fashion in America, but also to examine the appropriateness to Britain of the ideology upon which the American approach was based. Together with a study of similar thinking in Europe, particularly that of the Anglophile Austrian, Professor Hayek, a preacher of libertarian values long in the wilderness, the intellectual basis of the new right in Britain was established. It represented a combination of two powerful factors. The first was a willingness to listen to the libertarian wing of the Conservative party to a degree unprecedented in the party's history, and second, a willingness to assume that the American interpretation of the appropriate ideological underpinning of the market capitalist model applied equally to Britain. What emerged was a libertarian justification of the current economic and political system, the full practical consequences of which are still manifesting themselves. Where it has been given a religious expression it has closely reflected the Calvinistic tradition.

The ideological approach of this new libertarianism in economics can best be examined in three different ways; first, its implications for business ethics, i.e. the moral justification for various elements in business activity; second, its implications for the ethical content of the market place and third, its implications for the ideological basis for the political, economic (and, indeed, social) model generally.

Those engaged in business activity have always had to struggle against the poor image held of them by the rest of society. It was Adam Smith himself who suspected that when businessmen get together there is restraint of trade; this underlying hostility has permitted business, at best, to make a weak claim to work within a moral dimension. In contrast, many businessmen (and no less strongly than in Victorian days) would claim that they practise personal integrity in their daily dealings. Like those in any other area of human activity, they have standards of personal behaviour which mean that they do not lie, they do not resort to bribery, they respect relationships with others. Applying these principles in business life is no easier and no harder than anywhere else. It is also the case that many people in business have a high sense of calling over their work. The old fashioned term was the dignity of labour; that sense of a job well done is a major justification for work. It is, of course, often difficult to convey this attitude to others, especially when the precise content of their work appears mundane or even trivial. To say that one makes plastic ducks, for example, is no show-stopper at a cocktail party. However, anyone who has sat next to an irrepressibly buoyant American salesman on an early morning air journey in the United States, will quickly recognise the precise point that is being made.

Another area of business ethics relates to the behaviour of the corporation itself. There are many theories of effective management. Some of these depend on fear, either of bosses or of losing one's job, or they may be based on a crude view of motivation that is related only to cash reward; there are many cynical views of what makes an incentive. The American dictum, 'good guys come last' also has substance. There are other approaches that link effective performance to good management in an ethical sense. These models deal with matters like justice, what is good leadership, how is authority to be practised and what is the true nature of ambition and motivation. Much more will be said about these matters in a later chapter, but suffice to say that the increased interest in this debate, prompted by the ideological debate, is to be welcomed. Business ethics has also raised its profile in terms of corporate responsibility; how a corporation accepts responsibility for the well-being of the local community where it is located; whether it is concerned about global issues such as pollution; whether it accepts that stakeholders include others than simply shareholders or even employees. Corporations spend much time nowadays formulating mission statements that often embody an understanding of corporate values. All these elements of business ethics are important and it is good that there is now a context in which they are more openly and more fully debated.

An extension of this argument, that applies equally to individual businessmen and to corporations, is to tackle head on the issue of, what might be called, the morality of the market place. That is to say, to examine whether there is a moral content in the activity rather than simply to regard it as, for example, an area where greed is practised. Views along these lines have been well articulated in the United States (although Brian Griffiths [33] has argued the same broad case in the British context). Robert Wutherow, Professor of Sociology, Princetown University, for example, provides a typical argument [34]; he lists five ways in which the market can be claimed to be a moral activity.

First, the market requires the individual to act in community; his is not an isolated individual activity. On the contrary he works within a relationship with others, so it is a community activity which tests the rightness of his activities and the honesty of his deeds. Second, operating in the market creates duties. For example, a manufacturer will take on a duty to supply goods; that will create a contractual obligation which he must fulfil both for legal and moral reasons. Third, despite its imperfections, men and women are better engaged in trade than in, say, making war or stealing from each other. As de Montesquieu said, 'commerce polishes and softens barbaric ways'. [34] Fourth, the market offers protection against undue concentration of power. As most markets, certainly the ideal market, consists of many players, their activity puts a brake on those trying to gain overall control and provides a fence against the otherwise unbridled exercise of political power. The activity is therefore a means of providing peace and stability. Finally, and above all, participation in the market provides the opportunity for that most important human activity of all, the exercise of individual freedom, the acquisition of human dignity and the practice of self-reliance.

The extensive development of this interpretation of the ethical basis of market capitalism in British society today has been closely associated with the Conservative administration which has come into power in Britain since 1979. It is associated with the name of Margaret Thatcher but is also loosely related to the term monetarism.

The elevation of the latter term to such great significance has been curious particularly the way it, an arid technical term if ever there was one, has been given such moral content. When the term monetarism first reappeared in the early 1970s, it was employed in a narrow technical sense of commending Irving Fisher's monetary theory of inflation (encompassed in the famous identity $MV = PQ$). The theory had been around for a long time (Fisher described it in 1920) and, in the eyes of many, if not most economists, had been discarded. Its re-emergence caused no dramatic concern. Like their colleagues in the natural sciences, economists are always ready to debate the efficacy of any theory. It was thus seen as a re-opening of a technical debate, this time with the powerful new research of American academics such as Milton Friedman, to be called to its aid.

As that debate developed, however, it became clear that those people who were arguing the case for a monetarist approach to the control of inflation (a sensitive topic in the 1970s), could also be seen to be holding a purer view of market capitalism than was prevalent at that time. The advocacy of monetary policy, to control inflation, went hand in hand with the arguments for reducing state intervention, for freeing up markets, including the foreign exchanges, and for attacking vested interests wherever they were inhibiting competition; that wider debate embraced economic policies and issues generally.

However, further still along the debate, it became clear that those taking the side of this freer and purer market capitalist model, also held libertarian views about the role of the individual in society and what conditions had to be set to allow him or her to pursue his or her own route to life fulfilment. The view created a momentum that has provided a comprehensive rearticulation of the basis and role of most major social and social welfare institutions. The tools of this re-evaluation have been a market capitalist presumption, that wherever possible a market solution should be found (to ensure that choice was allowed to be expressed and that resources were allocated efficiently), and a libertarian view that people should be encouraged to be as self-reliant as possible - to take care of their own matters (and not depend on the state). In order to do so they should retain control of as much of their income as possible rather than pay in taxes to the government. This debate has resulted in a system now been put in place in Britain which is an outstanding example to the world of the application of a libertarian ideology in a modern complex industrialized society.

The well known and respected American advocate of the libertarian position, Michael Novak, has argued in similar terms; that capitalism must be understood not just as an economic system but as a triad of economic, political and moral-cultural systems. Each has its proper place. The task of the economic system is to create wealth but the economic system needs values from outside itself and for these it must turn to those provided by the moral-cultural system. [36]

Dykema, however, points out a weakness in this separation since, whilst each system must be 'faithful to its own mandate, it must likewise responsibly understand and reckon with the domain and needs of all the other spheres'. There is also the very real problem that many of the values to be found in the moral-cultural system 'find no particular honor in the market place. People motivated by concerns for the neighbor, for the poor ... frequently find themselves in tension with 'economic' values such as profit maximization'. [37]

This chapter has used a number of historical examples to make the point that each form of economic activity, economic organization, can be seen to be based on a particular ideology which, in the main, has been rooted in a particular religious or philosophical base. The forms of economic activity have changed; the ideological basis has also changed. However, it is very hard to avoid noting the paradox in the way in which changing values and the needs of expediency appear to be linked. At any one time people act with reference to a prevailing or preferred set of values to give them inspiration, to encourage them to renewed effort, and to provide a means of judging moral performance. The ideology provides the ultimate justification of their activities. Nowhere is this need for justification more evident than amongst those who are benefiting by exploiting new circumstances and new conditions. Nowhere is a moral basis for action seen in clearer terms.

It is equally clear that people appear to have no trouble in selecting a particular value set, as it were, to suit their ethical needs, to provide a justification particularly suited to the circumstances of the time. The fact that different sets of ethical options are continually on offer, creates circumstances of relative ethical choice and action, which clashes with the perceived need for absolute standards in guiding daily actions.

The strength of the need for ethical justification amongst the winners in the process of change, cannot be overestimated. It is often missed or allowed to go unheeded because the winners begin as a minority who barely challenge the vast majority sitting comfortably in the safe haven of the status quo. Where potential winners are seen to break down barriers, they are iconoclasts, upstarts, their new value sets are unfamiliar and frightening.

There are two sets of losers in this ethical game. First, there are those who were the previous generation of winners. Those who fought and won earlier battles and, as a result, imposed their own innovative values on society. The time since then had been spent in consolidation and refinement of the system developed to meet their needs as perfectly as possible; their reactions to new ideas and new values are purely defensive and they yield only slowly to the attack on what has now become conservative positions.

The second set of losers comprises a group that deserves far more study. It is the generality of people who, through history, have been on the receiving end of change and whose position, whose wants, whose preferred standards have always been less than clearly articulated, even in this modern day of the democratic process. The second group is a living witness to others in society and, from time to time, its presence will prick the conscience of the winners to a point where action, even against the self-interest of the winners, is provoked. More importantly, they (the second group) exert a continual pressure; they are a continual unnerving, uncertain, uncomprehended element in society that will demand to be heard eventually. The nature of the pressure exerted by this group of people has varied from period to period, but its pressure has always been there. In earlier days it was often expressed in populist terms through the customs and practices of the mob. Since the mid-nineteenth century its potential power has been there for politicians to court through the democratic process.

Notes

1. Robinson, Joan op. cit. p. 18
2. Tawney, R. H. (1990), *Religion and the Rise of Capitalism*, Penguin Books, Harmondsworth, p. 30
3. Quoted in Davies, G. (1959), *The Early Stuarts 1603-1660, Oxford History of England*, Oxford University Press, Oxford, p. 194
4. Quoted in Davies, G., ibid. p 129
5. Quoted in Davies, G., ibid. p 198
6. Quoted in Desmond, A. and Moore J. (1991), *Darwin*, Penguin, Harmondsworth, p. 5
7. Quoted in Hill, Christopher (1988), *A Turbulent, Seditious and Fractious People*, Clarendon Press, Oxford, p. 176
8. Tawney, R.H., op. cit. p. 34
9. Tawney, R.H., op. cit. p. 35
10. Tawney, R.H., op. cit. pp. 43-44
11. Tawney, R.H., op. cit. p. 44
12. Tawney, R.H., op. cit. p. 44
13. Tawney, R.H., op. cit. p. 45

14. See Tawney, R.H., op. cit. chapter 2 for a more detailed discussion
15. Tawney, R.H., op. cit. p. 111
16. Tawney, R.H., op. cit. p. 117
17. Tawney, R.H., op. cit. p. 117
18. Tawney, R.H., op. cit. p. 120
19. Tawney, R.H., op. cit. p. 121
20. The Bible, The Second Letter of Paul to the Thessalonians Ch. 3 v. 10
21. Tawney, R.H., op. cit. p. 252
22. Tawney, R.H., op. cit. p. 252
23. Quoted in Desmond and Moore, op. cit. pp. 394-5
24. Desmond and Moore, op. cit. pp. 420-1
25. The concern for the poor was not, of course, limited to Christians who took the socialist position. Those who worked to help the poor held all types of theological positions and did not necessarily embrace socialism.
26. The following paragraphs owe much to Dennis, N. and Halsey, A. in their excellent study (1988), *English Ethical Socialism*, Clarendon Press, Oxford
27. Tawney, R.H. (1982), *The Acquisitive Society*, Wheatsheaf Books, Brighton
28. Dennis and Halsey, op. cit.
29. Dennis and Halsey, op. cit. p. 200
30. Dennis and Halsey, op. cit. p. 201
31. See Preston, R. (1979), *Religion and the Persistence of Capitalism*, SCM Press, London, for a contemporary assessment of the relevance of Tawney's ideas to current church and society issues.
32. Weiner, Martin J. (1981), *English Culture and the Decline of the Industrial Spirit 1850-1980*, Cambridge University Press, Cambridge
33. Griffiths, B. (1982), *Morality and the Market Place*, Hodder and Stoughton, London
34. Wutherow, Robert, 'The moral crisis in American capitalism', *Harvard Business Review*, March-April 1982
35. Ibid.
36. See Dykema, E.R. (1989), *Wealth and Well-Being: The Bishops and Their Critics in Prophetic Visions and Economic Realities*, Strain, Charles R (ed.), Eerdenoms Publishing Company, Grand Rapids, Michigan, pp. 51/2
37. Ibid.

7 Social response to economic change

One interpretation of a heavy structure is an institution which, by intent or circumstance, impedes change. Alternatively, it can be regarded, more neutrally, as an institution incorporating an alternative value set demanding the right to co-exist with institutions based on other value sets. In the former case the task is to examine the pathology of, and reaction to, change. Those reactions may be entirely negative in the form of outright opposition to new systems, new ideas and new values. In the latter case, however, it may be necessary to deal with a quite different matter. Society, in its widest sense, contains subsets which have differing objectives, use differing means to achieve these objectives and are devoted to optimizing satisfaction based on different value sets. The task of exploring the impact of change is, then, to examine how different systems co-exist and how they balance each other. The circumstances may still be largely negative, in the sense that the process seeks to provide checks and balances which ensure that no one subset of society is allowed to go entirely on its own and ride roughshod over other equally valid subsets. However, it can be regarded in a much more positive light; the task being to ensure that different systems and value sets are blended together to deliver a satisfactory range of overall social activity; an optimization of society's objective function, to use a now unfashionable term.

As we have seen, when social change is taking place, the active stimulus or driving force is often of economic origin, exerting pressure to create conditions that allow it to break free and flourish fully. This pressure invariably results in clashes with countervailing forces in other parts of society. What is revealed, therefore, is essentially a reactive characteristic of the social structure. It is a structure as much on the move as the economic model itself but the movement is a constant attempt to re-establish itself in relation to economic change. The social structure's task is to provide society with balance, ensuring there is an appropriate reflection in those social structures of other value sets to those underpinning economic activity.

These protective and reactive systems cover a large range of social activity. They may be reflected simply in common custom and social courtesies (often expressed in a generally accepted sense of justice or fair play), values that are deeply embedded in society. However, as society has become more complicated, its defences have become institutionalized in forms such as the education and the legal systems and other institutional networks of society.

For most of British history the supportive (protective) social structure was the village. A social and economic network built round the village lasted for many centuries and was not seriously attacked, as the normal economic and social unit, until the late eighteenth century. The structure changed in detail over time, but continued to reflect the same essential ingredients. On an economic level, the primary activity was agriculture and its associated occupations, linked to processes of production such as blacksmiths, toolmakers, bakers, butchers, shoemakers, etc. The distribution of income from agriculture depended on access to land. Sometimes it was in the form of a landlord/tenant relationship, or at an earlier time contained an element of villeinage or tied labour, or it was related to access to open or common land.

The various forms of economic organization either reflected a social hierarchy or determined it; whichever was the case the economic status of the members of the village was an important element in its resulting social structure. The growing numbers of independent or semi-independent landowners behaved in a different way, and were treated in a different way, from those hired or tied by a wage basis to the larger landowners. For most of the time the social structure, associated with this form of economic organization, was intensely hierarchical and created firm barriers preventing movement between classes.

The hierarchy nevertheless carried with it equally firm social obligations. These were exercised on a personal basis, to fulfil the duties of class, especially on feast days; such obligations were, of course, commensurate with status. But there was also a communal obligation, by which the village as a whole accepted responsibility for its disadvantaged members. Under this system the deserving poor were adequately looked after. However, the undeserving poor of the village were treated harshly. The outsider in difficulty, who came into the village, received (as a vagrant) short shrift indeed.

Whatever the limitations of either social attitudes or the availability of resources to support the disadvantaged, the village system provided a social welfare net of sorts by custom. The rights and responsibilities that went with it were clearly understood and defended. These rights and obligations are reflected in the rich seams of British social history. For example, the annual beating of the bounds of a village reflected the community's declaration of its boundaries and extent of its jurisdiction and thus the land area from which the village, as a community, expected to earn its living. Another example was the adjudicating role played by the lord of the manor, who was invariably the justice of the peace for the village and its environs. The tradition of the English magistracy is an excellent example of the way in which a local social nexus evolved to provide support and protection to the local community as well as administer justice and maintain order.

The transitional period, between mediaeval life and the industrialization of the nineteenth century, was influenced by the transformation of traditional patterns of economic activity and the growth of state activity. The period represented a time in Britain's history when considerable state intervention was practised. At local level, this was reflected in the social control over market transactions imposed by price regulation (in respect of sales of products) and the regulatory system imposed by guilds (in respect of wages and working conditions). At local level, too, a protective social system was actively enforced. At national level, the preoccupation with the national interest and the use of regulatory legislation to promote that interest, also provided an institutional framework for addressing the well-being of the community as a whole.

The political ferment created by the commonwealth period, religious nonconformity and the associated development of political thought and practice, began to play a role in constraining economic power and its consequences. Whilst this ferment did not lead directly to the creation of radically new social welfare institutions, for it was not set in a period of rapid economic change, the increase in political awareness amongst common people was irreversible. Sources of modern British democracy can be found in this period which provided the basis for the development of the next two centuries. The agitators in the New Model Army, and religious leaders such as John Bunyan, were the founding fathers of a populist movement which exercised considerable political power in the years to come.

For preachers like John Bunyan, their duty was not only to call the faithful to pursue the painful path to salvation. It was also to demand freedom for the common people, to attack the aristocratic rule of the gentry and to condemn the lifestyles of the rich. In one sense, Bunyan, who rose from working class stock, appeared to be a simple evangelical preacher, but to many in authority his activities were seditious and, indeed, he was imprisoned for exactly that crime. The freedom to preach, which he struggled to claim all his life, was seen to be a close relative to subversive political agitation. He was the product of a radical, albeit defeated, revolution and in the rise and consolidation of dissenting interest, Bunyan became a national figure as a proponent of plebeian culture. [1] Whatever the judgement of Bunyan's influence as man and preacher, his writings, despite all attempts at censorship, were published and circulated massively. The popularity of *The Pilgrim's Progress,* both its strong religious appeal and its egalitarian political message, became and remains the bible, and textbook, for people seeking liberation both from their sins and from political oppression throughout the world.

Whilst the protests of the many, such as those to be found in Bunyan's congregations, appeared to be futile and unheard, the reality was different. As E. P. Thompson has illustrated in his fascinating study [2], British society has always contained a populist culture which, even if it lacked articulation, was a force to be reckoned with especially in the period beginning in the middle of the seventeenth century.

Thompson stresses the importance to populist culture of the traditions passed on by means of customs and which invariably provided the basis for resistance to change. He also illustrates how the activities of the mob and the use of riots were a means of expressing dissent and discontent. To many, the two terms 'mob' and 'riot' both carry seditious overtones, particularly to those in power even today, but Thompson shows that it is possible to take a much more relaxed view and to see them both as legitimate expressions of populist politics. Certainly at the time, and at local level, many leaders such as the Justices of the Peace, were fully aware of what was being said to them by the mob and were equally aware of the need to avoid or repair injustice to which the mob was drawing attention.

Custom amongst common people is a mixture of culture and folklore and practices legitimized by ages of use. The persistence of these practices brings them close to if not identical with rights. As Thompson puts it,

> Custom was certainly a 'good' word in the eighteenth century: England had long been priding herself on being Good and Old. It was also an operative word. If, along one path, 'custom' carried many of the meanings we assign now to 'culture', along another path custom has close affinities with the common law. This law was derived from the custom, or habitual uses, of the country: usages which might be reduced to rule of precedents, which in some circumstances were codified and might be enforceable at law. [3]

Thompson quotes as an early example the *lex loci,* the customs of the manor. 'These customs, whose record was sometimes only preserved in the memories of the aged, had legal effect, unless directly voided by statute law.' [4]

In the eighteenth century the plebeian culture reflected a number of characteristics. It was in great part made up of customs and traditions passed on from one generation to another, particularly within the family; household duties for the young daughter, child rearing for the young mother, employable skills for the son. This passing on of culture was essentially traditional and preservative, but there was also a rebellious element in traditional culture:

> The conservative culture of the plebs as often as not resists, in the name of custom, those economic rationalizations and innovations (such as

> enclosure, work discipline, unregulated 'free' markets in grain) which rulers, dealers, or employers seek to impose. Innovation is more evident at the top of society than below, but since this innovation is not some normless and neutral technological/sociological process ('modernization', 'rationalization') but is the innovation of capitalist process, it most often expressed by the plebs in the form of exploitation or the expropriation of customary use-rights, or the violent disruption of valued patterns of work and leisure. Hence the plebeian culture is rebellious, but rebellious in defence of custom. The customs defended are the peoples' own, and some of them are in fact based on rather recent assertions in practice. But when the people search for legitimations for protest, they often turn back to the paternalistic regulations of a more authoritarian society, and select from amongst these those parts most calculated to defend their present interests. [5]

Furthermore Thompson points out:

> In another sense the problems are different, perhaps more acute, for capitalist process and non-economic customary behaviour are in active and conscious conflict, as in resistance to new patterns of consumption ('needs'), or in resistance to technical innovations or work rationalizations which threaten to disrupt customary usage and, sometimes, the familial organization of productive roles. Hence we can read much eighteenth-century social history as a succession of confrontations between an innovative market economy and the customary moral economy of the plebs. [6]

This culture clearly anticipates what is to come:

> In these confrontations it is possible to see prefigurements of subsequent class formations and consciousness; and the fragmented debris of old patterns are revived and reintegrated within this emergent class consciousness. In one sense the plebeian culture is the people's own: it is a defence against the intrusions of gentry or clergy; it consolidates those customs which serve their own interests; the taverns are their own, the fairs are their own, rough music is among their own means of self regulation. [7]

This description of eighteenth century populist culture reveals a protective system that seeks to preserve many aspects of society against the impact of, mainly, economic changes. That protective system calls in aid, to justify itself, the customs of centuries past, which are by now embodied in values and require to be protected in some way against intrusion from alien values and circumstances.

When this line of analysis is pursued into the industrial revolution, two developments of major importance emerge. The first is the devastating impact on traditional life and customs that flowed from the disruption created by population movement and the exigencies of factory life. The second is the greater formalization of populist views, represented by the emergence of a distinctive working class and by the growth of democratic institutions.

As we have seen, the impact of economic change on English social life during the industrial revolution was massive. The over-riding feature was the massive movement of population off the land and out of the villages into the rapidly expanding towns and cities. This movement was accompanied by a shift imposed on the majority, into a wage earning, cash nexus basis, together with the imposition of monotonous work to replace the seasonal rhythm of rural life and the need to live and work in the big cities. The practical consequences were bad enough, smoke, filth, slums, disease, etc., but the transformation also destroyed traditional concepts of society and for the first time created real physical barriers between the rich and the poor. The latter were now excluded socially and geographically from the lives of the former. Anomie prevailed. E. J. Hobsbawm quotes a clergyman [8] describing Manchester:

> There is not a town in the world where the distance between the rich and the poor is so great or the barrier between them so difficult to be crossed. There is far less personal communication between the master cotton spinner and his workman, the calico printer and his blue-handed boys, between the master taylor and his apprentices, than there is between the Duke of Wellington and the humblest labourer on his estate.

The social effects of this revolution threatened to blow away the protective elements in pre-industrial society. However, although it was a long hard haul, the impending disaster evoked a new expression of populist consciousness backed increasingly by populist power through either industrial action (a concomitant of the growth in trade unions) or the ballot box. Amongst other things these conditions led to a formalization of class relationships, which, in one sense, replaced an earlier reliance on custom but in another represented a new articulation of working class culture. It found expression industrially and politically but it also developed a powerfully new social nexus:

> The middle class view of Friendly Societies was that they were rational forms of insurance. It clashed head on with the working class view, which also took them literally as communities of friends in a desert of individuals, who naturally spent their money on social gatherings, festivities, and the 'useless' fancy dress and ritual to which Oddfellows, Foresters and other 'Orders' which sprang up all over the north in the period after 1815 were so addicted. Similarly the irrationally expensive

> funerals and wakes on which labourers insisted as a traditional tribute to the dead and communal reaffirmation of the living were incomprehensible to a middle class which observed that those who liked them were often unable to pay for them. Yet the first benefit paid by a trade union and friendly society was almost invariably a funeral benefit. [9]

In the new institutional framework required by the industrial revolution a key role fell to education. An adequate debate about the origins of the present British educational system, its purpose and ideology, would spread well outside the confines of this analysis. Much of education has been, and is, to do with incalculating values and culture, as well as providing basic learning. Such values, especially at school level, have much to do with the growing up process offering pupils the means to make the transition from personal to social values. The economic imperatives have also always been present, i.e. the need for education to equip students with the ability and skills required to earn a living. The first few years of the work environment, indeed, are often seen as the route by which young people complete their journey into the full maturity of adulthood (one reason why unemployment can be so devastating to young people thus robbed of the conventional circumstances in which to 'grow up').

No-one would be naive enough to believe that the education system has developed wholly on altruistic grounds. Education for the masses only became a possibility when it was seen as a necessary part of the industrial process. The cynical view cannot be wholly dismissed either that schools were provided to keep children occupied whilst their parents worked, or to incalculate standards of hard work for future use in the labour force, and to provide the basic vocational skills vital to an increasingly industrialized society.

These issues still remain and lack nothing in force. What is important, however, is to accept that the education establishment, if such a term can be used, represents a heavy structure which may well at any time create a clash with the imperatives of the economic model. The conflict may be functional, if the economic model seeks to demand an alternative output from the education sector, or it may be ideological if the two systems are based on conflicting value sets.

Another social institution which embodies wider social values is the legal system. Indeed, in the sense that the legal system continues to rely on common law, it goes right to the heart of the matter. Common law builds into every day affairs criteria that embody age old custom, an instinctive sense of right and wrong, and justice, which is there constantly to test the applicability and validity of changes taking place. In this sense the legal system is the quintessential heavy structure. New ideas, proposed new actions, institutions, laws or whatever, always have this final barrier to overcome - a test against tradition, common practice and common sense. This is not to claim that common law has been set in concrete and remains unchanged and unchangeable. Far from it, common law, through interpretation, is always on the move, but the important distinction is that it takes, as its vantage point, a far broader, a more comprehensive, view of overall social values.

The same argument can be applied to many other institutions in society although the issue of competing and conflicting values may not be so clear cut. The domain of religion is one which has been considered in other chapters. There have been many examples in recent years of prophetic statements by the church, seeking to offer a critique of economic change. Two such instances have been the American Catholic Bishops' statement on the US economy [10] and the Church of England's publication *Faith in the City*. [11]

The cultural backcloth of society, literature, drama, music and the arts in general (including, especially today, the role of television), all contribute to the establishment, preservation of, or the introduction of new, value sets into society, any or all of which, may at some time find themselves in a defensive role as against other changes being pressed on society by other value sets.

All these elements can act as components of a protective system which, when it is appropriate, can be called upon to defend values or, in other circumstances, where they unduly resist change, need to be overcome. A common feature of these institutions is that they are, in the main, reactive. They seem to be cast in the role of defending certain ideologies; a role that stands in contrast to the more generally regarded active, dynamic economic model constantly seen to be creating new ideas, new opportunities and demanding new room for expansion, for doing things differently. Thus the economic model comes in conflict with existing institutions and engages in battle to remove, or amend, that which stands in the way. Other elements in society, seeking to preserve and protect, are often seen to ally themselves with the losers in the game of change, rather than the winners.

Nowhere has this been more true than in the treatment of the economically poor in society; for centuries this group has been the battleground between those holding conflicting value sets. The scriptural view that the poor are always with us is a statement of fact. So far in history the majority of people have lived near survival levels. Even in western capitalist economies, the poor make up a significant minority, judged by whatever measure of poverty is chosen. The text can also be used to endorse the view that whatever is done to eliminate absolute poverty, relative poverty will always remain. Some would want to interpret the text more provocatively still to imply that society will always contain some people who cannot or do not try to help themselves.

When the subject of the poor is raised the distinction between the deserving and the undeserving poor is often quickly made. There is an objective and generally accepted view of the deserving poor. No-one blames the physically handicapped person for being physically handicapped or if they cannot get work. No-one blames the Somalian peasant for not having food, drink or shelter. No-one blames those who suffer in a natural disaster or from a major illness. Society is willing to provide assistance to those needing help for these reasons without judgement. Equally, although less well remembered, those fortunately not in these categories, do not have to accept blame for the plight of these others.

The 'undeserving' poor are a different matter altogether. There is nothing objective in the use of the term. To say someone does not deserve to be helped, is to pass judgement on that person's own attempts to help themselves. If people are lazy, feckless, irresponsible, or unnecessarily ignorant, it is quite clear where the blame for the consequential predicament lies. Moreover, just as with the deserving poor so with the undeserving poor, the labels given to them absolve others from the need to do anything to help. If the poor are poor because they are lazy it is quite clear where the responsibility for putting matters right lies. It certainly does not rest with others.

The concept of self-reliance has always been a key element in judging personal human activity. A quotation from St Paul was referred to in earlier chapters, 'If any would not work, neither should he eat'. [12] In the context of a struggle for survival this challenge looms large. In a subsistence economy everybody has to pull their weight. It is why female babies or physically handicapped children are so unwelcome in many subsistence societies. Self-reliance has a certain significance in a subsistence economy related to survival. So much so that, paradoxically, many subsistence societies have developed networks to help the community as a whole survive, rather than leave the task entirely to the individual. In such societies what is expected of an individual - to pull his or her weight - is set in the context of community values and obligations.

The meaning given to the term in richer societies, i.e. those that have pulled away from subsistence level, is somewhat different. Self-reliance there carries with it a claim of ownership to the resources that have been generated by that individual effort. If wealth has been created by individual effort, ownership is undisputed. If, however, that wealth has been achieved with the help of others, or from resources such as land or capital created by a previous generation, a new dimension is added. Self-reliance becomes part of a debate about who owns the surplus and what degree of responsibility each has for the other when work is a collaborative effort. Whilst the strong may blame the weak for their lack of effort, the challenge as to why the strong are strong, has always been there. The Lollard cry 'When Adam delved and Eve span who was then the gentleman?' is always pertinent. The preference to rely on self-effort, both to avoid asking for charity and to provide a sense of personal satisfaction and achievement, has coloured the judgement of poverty since earliest times. And yet a sense of community, especially in subsistence societies, has also been strong. The biblical tradition gives strong support for that view too. In the Old Testament book of Ruth, there is the story of the farmer leaving the edges of his field unharvested for the benefit of the poor. Jewish tradition has many rules relating to the need to share with less fortunate and to help those fallen into trouble make a fresh start. The tradition of allowing a bankrupt person to retain at least his tools is one such practice.

Societies have always had to make room for the role of sharing. Whilst self-interest appears to dominate action, enlightened thinking stresses the interdependence of the community and the privileged position and duties of the strong. It is hard to see how a young person born in a western society, of gifted and wealthy parents, with a first class education and doors opening onto a potentially rewarding career, can legitimately claim a superiority based on effort over a young Somalian or, for that matter, someone born in the inner city of London. Good fortune, provided by the accidents of birth, geography, culture, parents, etc., far outweighs the claims of personal effort. Moreover, modern industrialized society's workings are now so complicated that, when someone becomes unemployed, it is hard to apportion blame between the worker's importunate behaviour and force majeure (say, the business cycle) over which he has no control.

The need for charity, i.e. offers to help others without passing judgement, seems to be a persistent feature of society. But charity is a curious thing. Sometimes it seems that the poor find charity easier to practise than the rich. Often it is easier to offer charity at a distance rather than closer to home. Mrs Jellyby's 'telescopic philanthropy', delightfully quoted by Robert Pinker [13] from Dickens' *Bleak House*, is the application of the law that, 'the further away the object of our compassion is the more intense will be the feelings of concern and obligation it evokes', admirably makes this point. Nonetheless, charity has to begin at home, so to speak, since a key choice in every society is the decision of how communities take care of less fortunate members. It would be good to think that the main stimulus for such concern is altruism. More likely, however, it will be found to be self-interest.

However, the practice of charity is usually constrained by geography. Where the unit providing charity was the village or parish the definition of who was eligible for help revolved around boundaries. Help was for those of the village or parish and not for the foreigner (the vagrant). There is much evidence that this attitude still prevails. It was evident in the way in which American communities, in the great depression of the 1930s, offered food to those migrating to look for work but urging them to move on through their particular community. It can be seen in reactions to the unemployed in Britain's northern cities in the 1990s, who migrate to live on state benefit in the plusher towns of the south.

A feature of charity in the past has been that the beneficiaries had no say in the matter. However, as the reasons and conditions why the poor are poor become more complex, and less obviously pointing the blame at them, the poor have claimed the right to be heard. In the present day they can be heard, and their influence on welfare policy exerted, through the democratic process.

Historically, British public policy towards the poor has largely retained the distinction between the deserving and the undeserving poor. The main initial legislation was created in the sixteenth century particularly in the Elizabethan Acts of 1536, 1597 and 1601. Under these Acts a clear distinction was made between the deserving and undeserving poor and the responsibility for both was placed on the parish. This legislation gave the destitute a legal claim to relief. Local justices of the peace constituted the Poor Law Authority. The system was administered by the parish. The justices had power to impose a poor rate on parishes. The result was an oppressive treatment of the poor. 'The different classes of the poor received different treatment. The aged and the sick were to be relieved in their homes. Pauper children were to be boarded out and, when old enough, apprenticed to a trade. Vagrants were to be sent to prison or to a house of correction and the genuine unemployed were given work.' [14]

The administration of the system came more and more under central government control, mainly to ensure that the system did not fall into abeyance through lack of enthusiasm for collecting the poor rates. However, the increasingly severe treatment of vagrants, at a time when the enclosure movement was throwing more and more labourers off the land, put severe pressures on the system not least in financial terms. The Speenhamland decision of 1795 to subsidize wages brought matters to a head by increasing the relief burden enormously (and, incidentally, allowing employers to keep wages low as they knew they would be made up out of public funds). The system also came under challenge later, as laissez-faire ideology took hold, leading some to advocate the abolition of the system entirely. Malthus, for instance, declared, 'no man has a right to subsistence when his labour will not freely purchase it'. [15]

The result was a new Poor Law Act in 1834, creating an administrative system embodying the laissez-faire ideology of the time. It was built around the doctrine of less eligibility; that is to say, if relief were to be given, the unemployed should not be paid more than the lowest wage of someone in work. The Act was designed to stigmatize the unemployed in this way and in many others. The destitute were required to enter workhouses, husbands and wives were separated, parents were separated from their children. It was an extremely repressive system and features of it lasted well into the twentieth century, with the fear of the workhouse still present amongst older folk today. There was no doubt that the 1834 Act reflected attitudes towards the deserving and undeserving poor consistent with the ideology of the time whilst at the same time it struggled to adapt a pre-industrial administrative framework to the needs and conditions of an industrialized society.

Although the Poor Law system set out in the Act of 1834 remained on the statute books until after the Second World War, the industrial, social and political developments after 1834 led to a general modification of the application of the rules. In particular, the severity of the workhouse regime was relaxed and outdoor relief was granted more freely. The challenges to the system came from a number of directions. The policy of stigmatizing the whole of a family quickly came under attack, on the grounds that wives and children could not be blamed for the shortcomings of the father; this led to pressure to give more outdoor relief so that at least the families could avoid entering the workhouse. Pressure also arose to deal differently with the elderly and the sick. Whether or not such people had entered the workhouse for good reasons, if they became infirm through old age or were incapacitated through illness, they presented a different social and moral problem. Even if they stayed in the workhouse they needed care and medical attention. Such problems became so extensive that the administrative responsibility of dealing with them was divided up. One over-riding Poor Law Authority responsibility, for pauper children, was transferred to the education committee of the local council, that for the sick and the elderly to the health committee. Already, by the end of the nineteenth century, these various concerns raised by the treatment of the poor, were being caught up in wider approaches to social and welfare issues such as public health, education and, in particular, in the growth of the hospital system and the administration of local government.

Gradually during this period the strict distinction between the deserving and undeserving poor broke down. Local authorities were becoming aware of the nature of unemployment. If it was not evident from what they saw around them, the new trade unions and reforming groups made sure that government, local and central, faced up to the realities of the situation. The introduction of wider democracy was also ensuring that other voices were able to be heard.

The rich period of social change outlined led to the creation of the modern comprehensive social welfare system in Britain (and parallelled in other western countries). That system comprises an institutional framework emerging out of Poor Law policy, but is to be interpreted more widely to include other areas of policy, such as industrial training designed to assist mobility in the workforce and to improve its quality, the health system, state insurance and pensions, and state housing.

Modern welfare policy is normally expected to address four main areas of concern; income maintenance, provision for the handicapped, provision for improving the quality of society, for example, through education and the arts, and issues relating to the cementing of the social fabric. Income maintenance is, perhaps, the thorniest problem since the policy is a direct descendant of the primary provisions of the old Poor Law legislation and concerns the issue of guaranteeing a minimum level of income. The relevance of the issues raised by the earlier Poor Law is reflected in the debate currently being carried on in Britain (and elsewhere) about the nature of unemployment benefit. Whether it should be offered for an indefinite period or limited to a specific period. Whether it should carry with it obligations to produce evidence of effort to search for work and how stringent those requirements should be. Whether school leavers should have a right to go directly onto unemployment benefit straight from school. Whether the unemployed should be required to do work of a community nature, and, lastly, at what level should benefits be set.

The strictness with which these criteria are applied will reflect whether a hard or soft attitude is being taken to the issue of deserving/undeserving poor. Whether the rules of unemployment benefit are tight or loose, there is always the additional issue of what benefits, if any, should be provided for those out of work but ineligible for unemployment benefit. The present practice of providing income support as a final safety net still provokes similar issues, namely who is eligible for it, at what level benefit should be set, whether it is given entirely in income benefit or whether some claims should be met by offering loans.

In the case of policy towards the handicapped the issues are less controversial because they are regarded as part of the deserving poor. Nevertheless there remain major issues of policy to be faced. The physically and mentally disadvantaged group have always fallen uneasily between two stools. Traditionally they have been regarded as a medical problem and often therefore institutionalized. But keeping people in institutions is an expensive practice. It also severely weakens the ties between the disadvantaged group and the community at large. On the other hand to de-institutionalize the approach often turns a perceived medical problem into a social problem and moves the 'problem' from behind closed doors, where it was conveniently forgotten, and brings it into the midst of the community.

Much of this has been happening in present society and it is difficult to know whether the prime motive is financial, i.e. it may be cheaper to get rid of the institutions, or professional, i.e. it is more appropriate to deal with matters in the community. Two examples amply illustrate the issues at stake. The first is to be found in the recent history of the main charitable organizations working with children. Most such organizations would have a similar story to tell as the NCH, a Methodist foundation which was founded 150 years ago as the National Children's Home and Orphanage. Its original service was, as its name suggested, the provision of institutional accommodation for children, mainly orphans. After a few decades, 'orphanage' was dropped from the title and it became the National Children's Home and recently, simply, the NCH in order to concentrate on the provision of family home-type accommodation, rather than institutions such as orphanages. More recently still, its emphasis has moved to providing services to support children in need in their home environment. Thus there has been a complete swing from offering an institutionalized solution to care for children, to working within the community and supporting children in the context of family life.

Present practice is to seek to move the mentally handicapped out of institutional care and into the community, to become more the responsibility of social rather than medical services. This client group is now visibly part of the community and figures, for good or ill, in daily community life. Amongst other consequences, which partly reflect the changing social climate but also results from the heightened awareness this greater visibility creates, is a greater emphasis on human rights questions. The rights of disadvantaged people as full citizens, which are highlighted by bringing them into the community, are now argued much more forcibly. Equally, issues relating to discrimination have come to the fore, such as equal job opportunities or equal access to public buildings. When forced to face up to these matters, particularly when prompted by militant groups, the public is far less comfortable about regarding them as the deserving poor, largely because they are being obliged to discuss the issues affecting the disadvantaged on terms dictated by the disadvantaged.

The broad field of education, formal schooling or access to various forms of vocational or higher education or in the wider cultural sense, also reflects the view that many of society's goals are non-material and depend upon improving the quality of life of the citizens. Policy initiatives therefore cover a wide range of cultural activities or popular movements, which may be made the responsibility of either central or local government. Often the policy will consist of providing subsidies for various national activities such as museums, art galleries, orchestras or opera houses. At local level, government may support many community activities such as youth clubs, local festivals or sports events.

A typical example arising out of the British tradition is the financial support given for adult education. The provision of subsidized evening classes, summer schools, etc., with the ultimate purpose of improving the quality of life of ordinary people, goes back into pre-industrial times. The early pioneers such as Wilberforce and Shaftesbury, at the end of the eighteenth century, blazed the trail for a range of activities related to the task of improving people and bettering society. A number of these associations were referred to in Chapter Two. They were followed, in the nineteenth century, by the initiatives of people such as Robert Owen and by many initiatives of the churches. The desire, for example, of the Methodist church to provide education to the working classes, through adult Sunday schools led eventually to the creation of the Primitive Methodist church (a splinter branch created largely to try to keep in touch with working class people). Many of these church initiatives, including, for example, the PSA (Pleasant Sunday Afternoons) movement eventually created secular versions, such as labour churches, but they all reflected the desire to bring education to the masses.

The later work in adult education, provided by the Workers Educational Association in which R. H. Tawney figured prominently, has already been noted. By the time of the 1914-18 war, such activity was firmly secularized and established as part of a wider social policy.

It is difficult to judge how deeply this aspect of social policy is rooted in society. Concern for formal education is well established as an essential element in improving the quality of society, a view strongly held in the libertarian United States. Whether the other forms of education activity, particularly those sponsored by the state, are as durable has to be questioned.

The final area of social policy is more one of attitude than a set of specific policies. It reflects the fact that what binds society together is self-interest rather than altruism. One example of this was the pressure for improvement in public health in the nineteenth century. It can be held cynically that how the poor live, what squalor they surround themselves with, is their problem and no concern of the better off. To some extent, that attitude is reflected, for example, in the fact that, within towns and cities, there are areas where the rich live and others where the poor live. Such elements of segregation, and there are many others besides those related to geography, are all very well but there are limits to the extent to which one part of society can immunize itself from another. In these modern times of universal suffrage and democracy it is practically impossible, no matter what aspect of life is chosen. But there are more immediate issues. Whilst the poor may live separately, and others are inclined to let them live as they please, if their living conditions become so insanitary that they pose a threat to public health, all will want to take action. Disease is no respecter of age, wisdom or wealth and it soon becomes in the self-interest of the rich to pay for public health measures; to change from open to closed sewers, to improve the quality of drinking water, arrange for waste disposal, etc.

Similarly whilst it may be argued that there is more lawlessness amongst the poorer section of the community, more family violence, more (drink related) public disorder, more petty theft, the chances of containing it in one particular section of the community are slim even with enhanced policing. It is more than likely that, in cases of public discontent, the 'mob' will cross borders to give vent to their anger on the persons and properties of those they envy. It becomes important, therefore, in their self-interest, for the rich to pay for things, housing improvement, more jobs, etc., in order to minimize the risk of such disorders. The well-being of one part of society depends on the well-being of the rest.

The evidence of this chapter has, hopefully, demonstrated that the social welfare nexus of a society is a heavy structure. The welfare network stands in many senses against the pressures, particularly economic pressures, to bring about change, to open up frontiers created by new technologies and new economic opportunities. Part of the strength of its position is that the system is largely based on alternative value sets to those underpinning economic activity. The social welfare system reflects and relates to the culture of the losers. It reflects the values upon which the community seeks to organize itself in the face of adversity. It concentrates on damage limitation, on care, on survival. The social welfare nexus also contains a strong appeal to tradition and custom. Whilst good and fair practices in many cases are enshrined in common law they do not have to be. Custom itself is a very powerful institutional force whether expressed in law or not. There are generally perceived right and wrong ways of treating people. There are fair and unfair actions and consequences. Views based on well tried ways of doing things are powerful and they often provide strong and resolute opposition to perceived new or alien kinds of change.

A pertinent example is the way in which conditions of employment are currently being transformed. There has never been an explicit commitment to lifelong employment in Britain but, nevertheless, there has been a presumption, an expectation, especially amongst white collar workers, of reasonably permanent and pensionable employment. Prospects of such jobs with such conditions of employment are now poor and reflect the changing demands of new technology. As a consequence an adjustment in perceptions of employment and to job security will be required. Change of such magnitude in working and career conditions, and its threat to 'custom', is likely to run into trouble, at least be slowed down, when people realize the extent of the change expected of them. The end period of a severe recession is not the best time for a protest to be articulated, when people are hungry for work, any work, but eventually society will want to address the conflict between the requirements of change and those perceptions embodying ideologies giving more weight to fairness, security, etc.

The earlier reflections about the ideology of what was known as the 'mob' in the eighteenth century, suggest the existence of a less formal heavy structure. Such terms as the mob are not often used to describe means of expressing protest today - certainly not legitimate means. However, the same conditions prevail. People in the mass have a disconcerting, but nonetheless effective, way of making their point. If they do not want to be led in a particular direction somehow the plans to make them will be thwarted. Anyone who has worked in manufacturing industry will have come across the bad manager who tries to bully or threaten his workforce to work harder only to be foiled on all sides in all manner of subtle ways - and rightly so. The point can be made more generally. When people at large think something is unfair they will find a way of successfully opposing it.

The social welfare system includes an important and coherent ideology embodying, broadly speaking, community values. However, it is striking that these values are deployed, and their strength is tested in practice, largely in a reactive way. That is to say, the circumstances for the debate about social policy appear usually to be created by economic activity. The latter, in the process of providing the gains enjoyed for many, seems to create adverse effects which push society to a point where community values based on justice come into play and demand change or at least a remedy or alleviation. This dialectic has been a powerful dynamic force in the development of modern western societies. What is surprising is that the relevance of the ideologies, upon which society's reaction is based, has so consistently been debated simply as reaction to excessive change created by economic circumstances. After all, economic activity is only part of wider societal activity and it would be reasonable to expect society to dictate the ideological context within which economic activity has to operate. Indeed it is possible to go further and argue, as will soon be done, that the conditions for economic success themselves depend on the ability of a society to create a way of carrying on economic activity based on values consistent with those held by society at large.

Notes

1. See Hill, Christopher, op. cit. p. 373, for comment on Bunyan's importance in this respect.
2. Thompson, E.P. (1993), *Customs in Common*, Penguin Books, Harmondsworth
3. Thompson, E.P., ibid. p. 3
4. Thompson, E.P., ibid. p. 4
5. Thompson, E.P., ibid. p. 9
6. Thompson, E.P., ibid. pp. 11-12
7. Thompson, E.P., ibid. p. 12
8. Quoted by Hobsbawm, E.J., op. cit. p. 87
9. Hobsbawm, E. J., ibid. p. 88
10. Economic Justice for All: Catholic Teaching and the US Economy. National Conference of Catholic Bishops, published in Origins: N.N. Documentary Service 16(2), November 1986
11. *Faith in the City*, (1985), Church House Publishing, London
12. The Bible, The Second Letter of Paul to the Thessalonians Ch. 3 v. 10
13. Pinker, Robert (1979), *The Idea of Welfare*, Heinemann, London, pp. 3-4
14. Birnie, A. (1948), *An Economic History of the British Isles*, Methuen, London, p. 131
15. Birnie A., op. cit. p. 377

The basic welfare state [illegible] an important and coherent ideal [illegible] community, broadly speaking communitarian values. However, it is striking that these values [illegible] and their strength [illegible] way. [illegible]

[illegible]

[illegible] society and [illegible]

Notes

1. [illegible] op. cit., p. 374 [illegible] importance in this respect.
2. Thompson, E.P. (1963), [illegible] Harmondsworth.
3. Thompson, E.P., ibid., p. [illegible]
4. Thompson, E.P., ibid., p. [illegible]
5. Thompson, E.P., [illegible]
6. [illegible] ibid. [illegible]
7. [illegible] ibid., p. [illegible]
8. Quoted by Hobsbawm, E.J., op. cit., p. 87.
9. Hobsbawm, E.J., ibid., p. 88.
10. Economic Justice for All: Catholic Teaching and the U.S. Economy, pastoral letter of the U.S. Catholic Bishops, published in Origins, N.C. Documentary Service, 1984, November 15.
11. [illegible] (1987), [illegible]
12. [illegible] Labour [illegible] in the [illegible]
13. [illegible] (19[illegible]), [illegible], London, pp. [illegible]
14. [illegible] (1948), [illegible] History of the [illegible], London.
15. [illegible] ibid., p. [illegible]

8 Working in community

An exploration of the ways in which ideology affects the way people relate together and, particularly, the way they engage in business activity together, needs to pay particular attention to the art of management. Management is one of the principal human relation skills; where people have to find ways of getting on together to their mutual satisfaction. Management normally carries with it implications of leading and being led. Sometimes that relationship is oppressive and expressed in a cold employer/employee relationship, with power resting with the employer using the threat of dismissal to ensure discipline and to set rewards. Sometimes that relationship is more open, more participatory.

In both extreme cases an ideology underpins the behaviour of the parties to the relationship. In both cases the primary issue is functional since the task of management (a word whose original root derives from the discipline and training of horses) is to ensure that jobs are done and purpose fulfilled.

The organization of production and supply in a modern complex industrial society, i.e. the role of management, is, therefore, a supreme example where there is likely to be a clash of values; where the choice of the appropriate value set may be crucial not only to satisfy conditions relating to the well being of society at large, but especially to ensure that the economic model performs efficiently.

This proposition will be tested by looking briefly at various forms of management that have existed in the past and then address the issues raised by modern management. A look into history will tell us much but not all. Much research has been done by economic historians to balance up the gains and losses of different groups and classes in society during the process of industrial development. What is lacking is substantial research into the precise forms of working relationships, by what principles they were governed and by what ways leadership was claimed or imposed. The lack of research may not be too serious when discussing pre-industrial forms of organization, where simpler systems prevailed, but it is serious when the industrial revolution is examined. The transition from a clear cut, direct master/servant relationship, to more modern forms of organization was as dramatic a change as the industrial revolution itself. It introduced two key elements, the institutionalization of employer/employee relationships and the need for new management techniques to cope with a variety of vital new issues. These issues included those facing organizations employing considerable numbers of employees and operating advanced capital goods. They also related to the task of organizing a number of different but interactive activities in the context of a competitive market. The result of the first was to set up formal clashes between owner and employee based on differing ideologies which, by being expressed in an adversarial form, made it difficult to search for grounds of mutual self-interest. The result of the second has been to build into the process of managing companies, an essential paradox between the view (internal to the company) that it is best to use a cooperative model to get the best out of people, whilst, externally, the activities of the business vis-à-vis other businesses are seen essentially as competitive - often expressed in the military terms of war.

The form of organization and type of management, if that is the appropriate term to use, of early, pre-industrial, farming is well documented. The British system of agriculture, sooner than most other European countries, left behind repressive elements of feudal life. By the eighteenth century a common pattern of mixed elements co-existed. These elements ranged widely but the commonest form was a village-based wage earning activity derived from farming.

The earlier form of villeinage, associated with the manor, gradually declined from the fourteenth century onwards. The growth in population, leading to the cultivation of much previously waste land, the growth of trade between country and town and the early use of cash rather than barter, all led to the emergence of a class of agricultural labourers as distinct from a landowning peasantry. These trends were reinforced by the consequences of the Black Death which reduced the working population and led to an increase in labourers' rights - although not without much friction between labourers and landowners.

However, even by 1750, the landed interests continued to dominate British life. The countryside was mainly owned by the larger landlords and was farmed by tenant farmers, using hired labourers. Hobsbawm [1] estimates that, 'by 1790 landlords owned perhaps three-quarters of cultivated land, occupying freeholders fifteen to twenty per cent, and a 'peasantry' in the usual sense of the word no longer existed'. In the majority of cases land was being farmed by a management organization consisting of a landowner (probably also the owner of whatever other capital was required - supplying seed and financing labour between crops) and wage earners. Discipline was undoubtedly harsh, especially once the enclosure movement set in and labour was plentiful, but it was a work regime as much dominated by the seasons and natural events as by any coherent management strategy. Also, the conduct of farming business took place within the strict and paternalistic social structure of a village/parish. One consequence was that the employer, the lord of the manor, would also, as a leading figure in the parish, probably be a magistrate and hold great power over many aspects of society, poor relief, punishment for crime, adjudication in civil matters, etc.

The system was harshly biased towards the ideology of those at the top of this rigid social hierarchy. Nevertheless, as the many rebellions and protests led by agricultural workers indicate, there was always protest and pressure to gain concessions. But the rules of the game were essentially dictated between social groups, each of which acknowledged its place, and by broad ethical principles prevailing within each class and group.

Anyone who is familiar with Elizabethan literature, such as Thomas Decker's *The Shoemakers' Holiday,* will have noted the camaraderie and joie de vivre with which the skilled craftsmen of the day are portrayed. Simon Eyre is the rumbustious master of his workers and apprentices, fiercely proud of their craft skills and their sense of community. Simon, himself, although only a craftsman, plays his part in the politics of the city of London and through the fortunate, but untimely deaths of some seven more senior aldermen, becomes Lord Mayor. Leadenhall market was built for him, and he was able to order that all shoemakers could close up shop every Shrove Tuesday. The picture is almost idyllic and contains essential elements about the way men and women related together in the skilled crafts that became important as towns developed. The sense of community was based to a large extent on the inter-relationship between craftsmen of the same craft. The relationship led ultimately to the emergence of the guild system itself. The guild became the focus of social activity, for political aspiration and for the regulation of working conditions, and control of the market. The community also continued to contain, within the context of a single craft workshop, a relationship between master and man as rigid and hierarchical as anything to be found in agriculture. Moreover, as the town industries expanded, the guild masters became even larger employers with the result that fewer craftsmen could look forward to the prospect of being their own master. Hence a growing class of journeymen emerged.

Both the pre-industrial agricultural scene and the mode of working in the towns in pre-industrial trades, indicated the prevalence of direct, hierarchical working relationships. The relationship reflected particular class positions, but it also reflected a close link between economic and social factors within the community. The sense of social order also contained an understanding of the role of work and its relationship to wider life and played a part in the social observing of rituals. Gradually pressures for change were brought to bear on the older guilds. With widening markets, craftsmen increasingly looked towards middlemen to market their goods. Also an increasing division of labour meant that craftsmen no longer made the complete article but was simply one element in a long line of production.

Another form of organization, which was in place at the time of the industrial revolution, was the domestic system. It was particularly common in the wool industry and was a system whereby a key individual (usually the clothier) was responsible for organizing the process of marketing clothing, by buying the raw material, supplying it to different groups of workers, to be worked up into the finished product, and ultimately marketing the finished product. Although the system was essentially a rural or small town activity it introduced the novel element of anonymity into contractual relationships that was to become such a common feature of industrial labour relations.

Amongst the various elements and characteristics of the industrial revolution none deserves to be described more than the transformation in the processes of production, and its consequence for the working lives of men and women. That which released the forces of change was the new role of capital, not in its monetary sense, though that was, also, a preoccupying issue, but in its physical sense as part of the means of production. Capital dictated a complex form of production which quickly led to the distancing of the owner of capital from his fellow workers. It transformed, initially in disagreeable ways, the conditions under which employees had to work.

Processes of manufacturing were already common by the middle of the eighteenth century, but the owner of a business still continued to combine the features of capitalist, entrepreneur, manager, and, often, worker. Mantoux [2] provides the illustration of the 'eminent manufacturer' in Manchester in 1720 who,

> ... would go down to his workshop at six o'clock in the morning, breakfast with his apprentices on oatmeal porridge, and then set to work with them. Having gone into business without capital, he earned his living from day to day, and if, after years of hard work, he managed to save a little money, he put it by and made no change in his daily habits. He rarely left his workshop or his shop and only drank wine once a year at Christmas time. His favourite pastime was to go of an evening, in company with others like himself, to an alehouse, where the custom was to spend fourpence on ale and a halfpenny on tobacco.

The separation between capitalist and worker grew quickly. The capitalist's interests, to keep wages as low as possible as the prime way to make profits, conflicted increasingly with the interests of his workers. The manager cum owner of a factory allied himself with the banker and the merchant whilst his own specialism, of organizing production, became more clearly defined. The price was a widening, and, eventually, an unbridgeable gulf between him and his employees in self-interest and function.

There are many features of the British industrial revolution that are unique simply because it was the first such revolution. One of these was the creation of a group of people specializing in the organization of production. The pioneers in this work emerged from all manner of places. Inexperience was rife and failures common. Mantoux [3] again quotes an extract from an 1803 enquiry into the state of the cotton industry,

> ...the following question was asked: 'Are manufacturers in general sufficiently acquainted with the process of weaving, to be able to determine a dispute which arises on the badness of the material?'. This was the answer: 'No, they are not capable of deciding those disputes that relate to bad materials: the reason of it is, that the master was never acquainted with the art of weaving. He just puts in a man who understands the trade, invests his capital, and when he gets the price of the market, he goes forward.'

Mantoux also makes the point that only rarely did the inventors of the new machinery turn into efficient and profitable manufacturers. Watt and Boulton were exceptions, but this was largely due to the business acumen of Boulton and, in particular, that of his supporting staff. But the majority of capitalists came from humbler less inventive stock.

> To succeed, these village agriculturists, blacksmiths, weavers and barbers, who made up the first generation of great English manufacturers, must have possessed in a high degree certain qualities fitting them for their new task, and these qualities, which they had in common, gave them a certain mutual likeness. Their distinguishing feature was not inventiveness, but a gift for turning other people's inventions to practical results. They were not all, like Arkwright, lucky or audacious enough to take complete possession of them and to secure a monopoly protected by patent rights. But, following the dictates of self-interest, they worked untiringly to reduce the inventor's legitimate rights to nothing. [4]

The principal skill required of early manufacturers was the ability to organize; to raise capital, to purchase machines and buildings, and then to recruit and manage labour. This latter task opened up the vast new area of industrial relations.

> Men used to working at home were generally not inclined to go to the factory. In the early days factory labour consisted of the most ill-assorted elements: country people driven from their villages by the growth of large estates, disbanded soldiers, paupers, the scum of every class and of every occupation. All these unskilled men, unused to collective work, had to be taught, trained, and above all disciplined, by the manufacturer. He had, so to speak, to turn them into a human machine, as regular in its working, as

> accurate in its movements, and as exactly combined for a single purpose, as the mechanism of wood and metal to which they became accessory. Hard-and-fast rules replaced the freedom of the small workshops. Work started, meals were eaten and worked stopped at fixed hours, notified by the ringing of a bell. Within the factory each had his allotted place and his strictly defined and invariable duty. Every one had to work steadily and without stopping, under the vigilant eye of a foreman, who secured obedience by means of fines or dismissals, and sometimes by more brutal forms of coercion. [5]

An excellent example of an owner who came to master the new skills was Arkwright, who quickly became well known for the discipline he established in his mills, although by the standards of the day he treated his workers reasonably well.

Another aspect of the new capitalism, which the smaller manufacturer had not had to face before, was the task of marketing. In pre-industrial times marketing was mainly a question of transporting goods to the nearest town and practising the skills required to trade in a local market. The history of, say, the Soho works of Boulton and Watt, reveals how important and necessary it became to create a selling network and eventually a sales force. The vital contribution that their loyal agent, William Murdock, made to selling throughout Britain, is well known. But the factory also began taking orders from Europe and as far afield as Russia.

The challenges to the early manufacturers were thus considerable and many of them barely rose to the challenge, either through ignorance, lack of experience, or lack of ability. Many of them did not inspire confidence in their contemporaries. Robert Owen wrote in his autobiography, 'The manufacturers were generally plodding men of business, with little knowledge and limited ideas, except in their own immediate circle of occupation'. [6]

There were others, however, of quite different character and quality. Notable examples were Matthew Boulton and Josiah Wedgwood. They were part of a community of men who, as Mantoux put it, realized

> ... some of the products of industry, pottery, for instance, are not only useful things. They have, or at any rate can have, an artistic value as well. A few manufacturers realized this, and their conception of their own occupations was thereby widened and changed. For them industry ceased to be only the means of acquiring wealth and power. When they tried to improve their equipment, or their methods of production, it was not only to gain a victory over their less conscientious or less careful competitors. It was also because technical progress, bound up as it was with the development of science and art, seemed to them a desirable end in itself.

Such purposes, higher than those of the mass of their rivals, did much to ennoble their lives and characters. [7]

Matthew Boulton was not only an efficient engineer and manufacturer, he was a cultured man who claimed amongst his friends Darwin, Priestley and Sir Joseph Banks, President of the Royal Society. He was a generous man who worked assiduously for the poor in Birmingham. He was deeply interested in inventions and in the new study of economics. He was also concerned about the ethics of business. As Mantoux relates

> His favourite maxim was Poor Richard's optimistic motto, 'Honesty is the best policy.' With reference to an agreement to be concluded with some of the firm's customers he wrote to his partner James Watt: 'You must not be too rigid in fixing the dates of payment. A hard bargain is a bad bargain. Patience and candour should mark all our actions, as well as firmness in being just to ourselves and others.' He completed his children's liberal education by advice founded on lofty moral principles: 'Remember I do not wish you to be polite at the expense of honour, truth, sincerity, and honesty, for these are the props of a manly character, and without them politeness is mean and deceitful. Therefore be always tenacious of your honour. Be honest, just and benevolent, even when it appears difficult to be so. I say, cherish those principles, and guard them as sacred treasures.' And he taught them not by advice only, but by the example of his own life. [8]

Josiah Wedgwood, too, was both a man of business and an artist. Although self-educated he was well read and had closely studied the art of antiquity. He was also an open-minded man, both in terms of scientific and philosophical enquiry, and in religious belief.

But these men were exceptional in the early stages of the industrial revolution. The majority were 'not like them in their finer qualities. They deserved admiration for their initiative and activity, their power of organization and their gift of leadership. But their one aim was money, men and things alike were only tools for the attainment of this single object'. [9]

Against this early owner cum manufacturer's view has to be placed the emerging view of those who were forced reluctantly into the new work environment; to be the subjects of the dictates of capitalistic factory production and an iron-willed managerial discipline. The contrast in conditions brought about by the factory, compared with the life of those already working in some form of manufacturing, was immense. The factory was literally hated.

> The feeling of repulsion which it aroused is easily understood, as, to a man used to working at home, or in a small workshop, factory discipline was intolerable. Even though at home he had to work long hours to make

up for the lowness of his wage, yet he could begin and stop at will, and without regular hours. He could divide up the work as he chose, come and go, rest for a moment, and even, if he chose, be idle for days together. Even if he worked in the master-manufacturer's house, his freedom, though less complete, was still fairly great. He did not feel that there was an impassable gulf between himself and his employer, and their relations still retained something of a personal character. He was not bound by hard and fast regulations, as relentless and as devoid of sympathy as the machinery itself. He saw little difference between going to a factory and entering a barracks or a prison. This is why the first generation of manufacturers often found real difficulty in obtaining labour. [10]

The harshness of factory discipline is well documented.

Discipline was savage, if the word discipline can be applied to such indescribable brutality, and sometimes such refined cruelty, as was exercised at will on defenceless creatures. The well-known catalogue of the sufferings of the factory apprentice, Robert Blincoe, makes one sick with horror. At Lowdham (near Nottingham), whither he was sent in 1799 with a batch of about eighty other boys and girls, they were only whipped. It is true that the whip was in use from morning till night, not only as a punishment for the slightest fault, but also to stimulate industry and to keep them awake when they were dropping with weariness. But at the factory at Litton matters were very different. There, the employer, one Ellice Needham, hit the children with his fists and with a riding whip, he kicked them, and one of his little attentions was to pinch their ears until his nails met through the flesh. The foremen were even worse, and one of them, Robert Woodward, used to devise the most ingenious tortures. It was he who was responsible for such inventions as hanging Blincoe up by his wrists over a machine at work, so that he was obliged to keep his knees bent up, making him work almost naked in winter, with heavy weights on his shoulders, and filing down his teeth. [11]

Even a little later when conditions were becoming more humanized the severity of disciplinary procedures offends the modern mind. [12]

So strict are the instructions [it was said of John Marshall's flax mills in 1821] that if an overseer of a room be found talking to any person in the mill during working hours he is dismissed immediately - two or more overseers are employed in each room, if one be found a yard out of his ground he is discharged ... everyone, manager, overseers, mechanics, oilers, spreaders, spinners and reelers, have their particular duty pointed out to them, and if they transgress, they are instantly turned off as unfit for their situation.

Or as Pollard describes conditions in the mining town of Neath. [13]

> ... most workmen were bound for fourteen years and a most severe set of rules and penalties regulated the men's lives. In the smelt house, for example, 'Swearing, cursing, quarrelling, being drunk or neglecting divine service on Sunday', were fined one shilling. Absence for two hours was fined by a day's wage, absence for a day by a week's wage, the fines to form the sick fund. Disclosure of secrets, fraud or mismanagement were to be punished by a fine of £100 - a sum not likely to be available to the ordinary workman.

This is a picture of conditions which revealed the absolute and uncontrolled power of the capitalist. [14]

> In this, the heroic age of great undertakings, it [the capitalist's power] was acknowledged, admitted and even proclaimed with brutal candour. It was the employer's own business, he did as he chose, and did not consider that any other justification of his conduct was necessary. He owed his employees wages, and, once those were paid, the men had no further claim on him: put shortly, this was the attitude of the employer as to his rights and his duties.

Given the speed and extent of change in manufacturing methods and organization that occurred during the later decades of the eighteenth century, it is surprising that an interest in, and concern for, what would now be called management studies, remained so low for so long. It was not until the middle of the nineteenth century that the methods used to control and manage the activities of an enterprise began to resemble modern practices.

Pollard [15] suggests three reasons why this was so. First, for a long time 'it was difficult to isolate the 'managerial' function from that of technical supervision or commercial control, which were often more critical'. Second,

> ... in a period dominated largely by pioneers and founder-managers apt to stress the differences in individual 'character', rather than the similarities generalized in a science of psychology, such questions as the structuring and management of firms must have seemed too individual, too unclassifiable, to repay further generalized study.

Third, there were two differing views about the treatment of labour.

> ... the views of the majority were bounded by the realization that they were dealing with a recalcitrant, hostile working force whose morale, whose habits of work and whose culture had to be broken in order to fit them for a form of employment in which they had to become obedient servants of the machine, of its owners and of crude monetary incentives.... In contrast with this we have the other pole, represented by Robert Owen

and his well-known plea to his fellow industrialists to treat their labour no worse than their machinery.

Until 1830, therefore, the former harsher attitude was dominant and in many ways fuelled the dialectical class struggle, which became the hallmark of industrial relations for over a century.

The first quarter of the nineteenth century can be taken as the low point when factory conditions were at their worst and ignorance of what needed to be done to control and manage the new system was most lacking. Developments, thereafter, steadily redressed the balance. The principal external pressures were political. Some of these were specific and related to legislation designed to curb excessive or harsh practices, limiting hours of work, improving factory safety, relaxing the stern administration of the Poor Law system. Some pressures were more widely political, such as the emergence of trade unionism and the development of a philosophy more inclined to accept state intervention as a necessary means of control.

At the same time factors internal to the business sector were also having their effect. Skills and techniques of management were all improving; accounting methods, management techniques of selling, ways of organizing production and of managing the labour force more positively through education and incentives were given. As these activities developed, in the context of the growing complexity and size of businesses, the gap between the owner cum entrepreneur and the other parties to the activity, grew larger and larger and was filled by the rise of a management class.

Pollard comments that management functions 'were essentially new during the industrial revolution, their novelty deriving directly from the novelty of large scale centrally controlled industry itself'. [16] But they played an increasingly important role since, 'management was the instrument, if not the originator, of many of the forces which shape the peculiar forms of rational self-interest, the ambiguities of ownership and control of firms, and above all the class relations within industry, which became peculiar to western industrial capitalism'. [17]

The same intrinsic polarization observed in the early days of the industrial revolution remains in place. On a mundane level all who have worked in industry can witness to the fact that two sets of values continue to co-exist. There is abundant evidence of the adversarial tension between employer and employee; yet at the same time there is a sense of community in the workplace. Both attitudes are there; modern management's task is to find a way of blending the essential ingredients of both to create the conditions for success.

Much has been written about management matters, especially during this century. No discipline flowered so quickly or so profusely and there are many, many theories and prescriptions, the detailed study of which must not deflect us from our main concern. McGregor's [18] classic distinction between theory X (the harsh) and theory Y (the benign) highlights the essential distinction between the two extremes lying at the heart of all management models. The propositions upon which theory X are based are as follows, [19]

1. The average human being has an inherent dislike of work and will avoid it if he can.

2. Because of this human characteristic of dislike of work, most people must be coerced, controlled, directed, threatened with punishment to get them to put forth adequate effort toward the achievement of organizational objectives.

3. The average human being prefers to be directed, wishes to avoid responsibility, has relatively little ambition, wants security above all.

The propositions upon which theory Y is based are [20]

1. The expenditure of physical and mental effort in work is as natural as play or rest.

2. External control and the threat of punishment are not the only means for bringing about effort toward organizational objectives. Man will exercise self-direction and self-control in the service of objectives to which he is committed.

3. Commitment to objectives is a function of the rewards associated with their achievement.

4. The average human being learns, under proper conditions, not only to accept but to seek responsibility.

5. The capacity to exercise a relatively high degree of imagination, ingenuity and creativity in the solution of organizational problems is widely, not narrowly, distributed in the population.

6. Under the conditions of modern industrial life, the intellectual potentialities of the average human being are only partially utilized.

There are many other models of management theory. Most attempt to address management objectively assuming the choices are value neutral. For example, management by objective or zero plan budgeting are presented as techniques designed to achieve a purpose, but they lack an essential ingredient if they fail to address the principal issue of how to encourage people to work together in community. There is much to be said about this both from the management and the worker's point of view.

People who go into management are motivated by altruism no less or no more than others. Management is a job, a career, an activity for which managers expect to be rewarded financially. For most of the time managers expect to enjoy their work. There are a number of elements against which managers would wish to judge their degree of job satisfaction. They include the sense that they are achieving objectives, that they are being allowed to pursue satisfactory means to achieve their objectives and that they are sufficiently rewarded.

For those who have worked in senior levels of industry the setting of objectives plays a key role - at least in terms of the internal operations of a company. The perception of the task of management is one in which the task is to achieve self-imposed targets for the business in the context of a hostile environment, which consists of a market or a situation, which is poorly understood because much necessary information is lacking, and where there may be competitors whose future tactics are less than perfectly perceived. There will also be random and unforeseen events and government intervention, often unpredictable and often hostile. Even when friendly government interventions occur these are often ineffectual. It is not surprising in these circumstances, to find management attitudes prevailing which are akin to those used by the military. Teamwork under threat to achieve specific targets, where failure means 'death', would be seen by many managers as an acceptable definition of management activity. It should not be assumed that working under these conditions is poorly regarded; on the contrary there is an excitement, a sense of fulfilment and achievement that can be intensely satisfying.

For most managers, much of such objective setting and planning is set in the context of the company's own internal activities. Managers work to demanding targets for sales, involving inordinate amounts of overseas travelling, or many hours of fruitless knocking on doors. They aim for production targets where maintaining capacity output is an overwhelming task; such people need little else by way of objectives.

Few managers would or should concede that their jobs are less interesting, less stimulating or less rewarding than others in society and in this narrow sense they probably rest satisfied with the objectives of their professional activity. The trouble is that for a professional vocation to be fully satisfying it requires to attract a sufficient degree of esteem from other parts of society. The activity needs to attract social value. Unfortunately this wider esteem is singularly lacking in Britain. For reasons much written about but still little understood [21], the people that make up the management cadre in industry, the engineers, accountants, salesmen, etc., appear to be poorly regarded by the rest of society in Britain. Because of this the justification for being a manager - at least the one given to and perceived of by society at large - deteriorates to being simply a way of earning money. Earning money (and maybe a lot of it) is seen as the principal reason why people become managers and is perceived as compensation for choosing not to pursue a 'good', socially highly valued job. Unfortunately, many managers have given up the attempt to refute this view and now act as if that is actually the case.

This low esteem is extremely damaging because it means there is little appreciation of the social role of 'industry'; as a major social agent affecting peoples' lives and whose contribution to the wider dimensions of society deserves to be given weight alongside the contributions of other social agents. If society does not recognise, and if managers do not take pride in, the value of the wider objectives of business activity, that activity will not be conducted in a way likely to assure economic success. Many goals in western societies are related to affluence. Affluence enables all manner of material wants to be met. It follows, then, that it is essential to acknowledge and praise those who create wealth and to give social recognition to those who have the heavy task of meeting these objectives.

It is commonplace to define the responsibilities of a company's management in relation to the stakeholders in the company, i.e. its legal owners, its customers, its workforce and society generally. In practice, the responsibility is often limited to the first of them - its legal owners; this is damaging since these objectives are essentially indivisible. Certainly to insist on including in objectives the task of satisfying all stakeholders' interests helps considerably to elevate the motivation of managers. They will be better managers for being encouraged to contemplate wider horizons.

One of the principal grounds for satisfaction in working in industry, certainly in management, is that it gives scope to the exercise of particular human skills and attributes. On the whole these are the more active, the more virile, the more competitive characteristics of the human personality. The first among many of these attributes is the ability to take risk - the desire to take on the unknown, the gamble with external circumstances. The art of risk-taking itself, accompanied by the decisive effects of winning or losing, carries with it many other activities which are equally rewarding and stimulating: the use of advanced analytical techniques, such as planning against uncertainty, decision-taking with limited time. The application of this wide range of skills is often an extremely satisfying experience.

The other side of the coin is that most managers are exposed, far more than most, to personal stress in one form or another. To a large extent this is inherent in a job which rewards success but punishes failure. As personal survival, as well as company performance, is often at stake managers are particularly in need of help for the personal management of stress, for coming to terms with the mobility and vulnerability attached to their careers. Not the least of these problems is the stress put on the manager's marital and family situation.

It is necessary to come back again to the fundamental problem, which is probably more society's than managers', that society (and certainly those concerned with values) holds an ambiguous view of the value of this set of human attributes. Is it best practice for humans to apply these heroic and aggressive attributes? Would it not be better if society relied less and less on these attributes and cultivated, instead, the more passive skills of reflection, contemplation, self-denial etc? There is certainly a case to be argued but the point that must not be forgotten is how much present standards of living owe to the way in which human endeavour has been channelled into industry. Assuming society continues to want more and more from its productive machine, which seems a reasonable assumption, it is a very risky business, and certainly inconsistent, to deprecate the very qualities that have made that possible in the past.

The foregoing argument does not mean that managers are simply risk-taking, money-making extroverts, singularly lacking in all human skills. It is simply not true of them as individuals nor is it an adequate listing of the skills required to make a successful manager and a successful company. The workplace situation in industry, the prevailing leadership model, is a practical example of human concern for others - where it really matters in a real situation. There is a view of industrial leadership that is seen to be harsh and uncaring, but in reality running a company is all about managing people. At the end of the day, whether things happen properly in a company will depend on whether people are getting on together, whether proper leadership is being claimed and real authority being given.

In some respects managerial skills are competitive and individualistic, but for success, such skills invariably have to be applied in a team context. The keynote of a good team is cooperation and the provision of mutual support, which will always be found in a successful business enterprise. This is a further example of the way in which good management can be regarded as an exemplary means of life fulfilment - by offering an opportunity for men and women to work together. One of the important challenges facing Britain, in particular, is to find a way of producing a proper blend of individualism and cooperation on which to base successful management.

It is almost trite to say that where a situation requires people to work together - which is commonplace in industry - the exercise of good human relationships is essential. Bully boys may have a passing effect, but (as any factory manager knows) they are eventually found wanting. Senior managers can create fear through threat of loss of jobs or lack of promotion, but the consequence is a demoralized team where initiative is left to others, where performance is jaded and demotivated.

The excitement in the community that makes up a business comes when leadership is being exercised correctly. Where, without any diminution in objectives, there is respect between all members of the team, all are taken into confidence, no-one asked to do more than the asker himself or herself is willing to do. Out of this atmosphere comes an excitement and a performance that produces results few economic textbooks provide for in their theory. A good manager can rightly claim to stand equal with those working in conventionally termed caring occupations in terms of leading people, respecting them and helping them where necessary; above all, in helping them to live a full life in relationship with their co-workers.

One final word about means. Much has been said in praise of managers, of recognising the qualities they exercise, and the need to reward them adequately. It needs to be said that this elevation of role of managers carries with it a concomitant heightening of the responsibilities of managers to the society they serve. Properly carried out management activity plays a vital role in cementing society together; that is why it is always regrettable when events suggest that managers simply act in a self-seeking, money and power grabbing way. It is equally regrettable if this key group is seen to identify itself politically too strongly with one particular set of (right wing) ideologies. Such a position not only belies their essentially consensus seeking role in society, but encourages workforces to take up countervailing ideological positions.

All want to be fairly rewarded for work done. To be rewarded is to be given a sense of satisfaction, a sense of achievement, derived from the recognition of others. Self-motivators (that blessed minority) can find this satisfaction from internal sources, but the vast majority require some acknowledgement from others to be convinced that they are doing desirable and socially acceptable things; this is why paid, rather than unpaid employment, is so important [22]. Being paid for work done is unambiguous. The payer must be satisfied with what has been done otherwise he would not pay for it. So it is reasonable to assume that money is the key form of reward required in industry; this is certainly the prevailing view in Britain. It has become fashionable to say that industry has not been paying its managers well enough. A better strategy, it is argued, is to employ fewer managers, but pay those retained much more. There is a place for this debate within the company. By and large managers look to their pay cheque as the prime measure of the company's assessment of their value. There is certainly much to be said for the annual face-to-face conversation over performance ending in reward (increase in salary) or not.

Whether money is that important is another matter. It is clear that at the highest levels of management money is not the prime motivation. A managing director drawing £500,000 per year must be close to satisfying his or her material needs (although the power of greed never ceases to surprise).

Most of the job satisfaction in these cases must lie in the exercise of power and leadership, and in the excitement of risk-taking. Further down the scale money may become more important, but the importance of non-monetary factors must not be forgotten - the need to feel that people are being treated fairly, the ambition to be exploited to one's full potential regardless of cash reward, and the existence of an open style of management. These key motivators in business are ignored at a company's peril.

It can be argued that everyone in industry, in a company, is a manager to a greater or lesser degree - even the lowest paid worker doing the most trivial task has to manage himself. If that truth were generally recognized it would lead to a quite different management model. This view can be found in some businesses, particularly non-unionized ones, and often in the United States. But it is not true in Britain in the overwhelming majority of cases. The historical reasons why such an approach is the exception rather than the rule, have been rehearsed earlier in this chapter. The reason the rank and file of most British companies do not see their relationship to the business in the same way as a manager does, has its roots in history, in the origin and development of the industrial revolution, during which employers built up great power over individual employees. This unequal balance of power was reflected in the setting of levels of pay and conditions of work. It was also reflected in the 'management model' in the sense that the relationship between employer and employee was perceived to be solely contractual; workers were engaged to do a job at a certain rate of pay (which the concept of the market system justified as 'good' for both employee and employer alike), with little of the modern perception of employee participation in the activities of the company.

It is clearly incontestable that in the early days the system was manifestly loaded in favour of the employer against the individual employee. Moreover, it would be foolish not to acknowledge the truth of much of the Marxist (and less extreme left) view about this situation - the degree of exploitation involved, the need to see the fight for justice as a class struggle, the central position of the debate over the right to property and the income therefrom.

The principal effect of this early period of exploitation was to create a counter movement to employer power: to organize labour, first through trade unions and second, in Britain, through a political party.

Much of the work of trade unions was devoted to improving the position of individual employees by negotiating minimum pay levels and 'fair' wage rises, and in arguing for improved conditions of work: workplace safety, workplace amenities, length of working week, paid holidays, and pensions. The defence of jobs was also gradually elevated in importance as an objective. In contrast worker interest in the running of a company, has always been minimal. The slogan, 'it is for managers to manage' has held sway for a long time and is still at the heart of the matter, despite recent developments and innovations. In most cases the union argument for workers' representation on company boards or in support of other forms of participation, has been put forward as much to increase bargaining power, by gaining better knowledge of companies' future plans and actions, as to help run companies better. However, whilst workers' interest in managing the company has been weak, their interest in the wider social structure has increased dramatically. The enaction of welfare policies, resulting in the provision of universal public education and a state health and social welfare system, is now seen as a proprietary issue. It is 'our' system, say the workers. Power has been gained through the democratic system - that power has given access to facilities that hitherto had only been available to those with money. In this sense the wider and narrower interests are joined; the maintenance of the welfare system became part of the industrial bargaining system in its widest sense. [23]

The present industrial structures owe much to past experience and for that reason represent a particular form of heavy structure, which may be hard to move. The well known historical factors leading to Britain's present union pattern, have resulted in the virtual absence of the concept of a company union in which workers' interests are identified with the future of one particular company. Rather, it is a system in which negotiations are conducted across a group of companies, or even nationwide, by industry or groups of industries. Whatever the gains of such a system, what is lost is any ability to relate rewards to specific company situations. In the circumstances bargaining 'deteriorates' to negotiating on the basis of wider generalities such as the level of national inflation, overall industrial productivity, and profit levels, or over a general cross union demand for shorter hours or lower ages of retirement. These issues are not unimportant, of course, and to a degree need to be addressed at national level, but unless the prime source of wealth creation, the growth in value added of an individual company, is at the centre of the stage, the key and prior issue for all debate about income distribution will be left out.

A second consequence, that flows from the historical context, is that the normal promotional scale of trade union leaders is from office at local factory/company level to regional and thence to national office. From the beginning, the loyalty of a potential trade union leader does not rest with an individual company. His or her eyes are set on a wider national stage and there is no reason, even at the lowest level of offices, to identify closely with a specific company. Apart from stamping industrial relations issues immediately as 'non-company', it means that an important source of leadership potential for the company is denied. In contrast, in Japan, holding trade union office in a company is seen as a stepping stone to senior management posts in that company. In Britain prospects are not seen in that light by trade unionists, nor, worse still, is it the attitude of management, who appear to regard any employee who opts to serve his union as suspect.

The sense of deep ambiguity over loyalty can be found amongst the rank and file workers in their company. Some of the statements made in recent years, by union leaders, in the context of mass redundancies, have been revealing. Take, for example, the growth in the idea that workers are 'stewards' for their job; that they must not give their job away (in return for redundancy payments?); that it must be held on to in order to pass it on to sons and daughters of the next generation. Or, to take another example, the controversy in the union claim that there will always be industrial conflict as long as managers 'had the key to the factory door' (and could lock the workers out); 'After all, that factory', say union leaders, 'is as much ours as theirs; we live and die in there'.

The arguments in both these examples are flawed in a strict technical sense. A concept of stewardship cannot be applied when the economic system lives by the laws of supply and demand. Management insists on the freedom to respond to market forces, which freedom has to include the freedom to create or destroy jobs. Similarly, in the case of ownership, it is just not true that workers own the assets of a private company. Under law, ownership is vested in the shareholders.

Yet the two examples touch on deep issues that relate to what is meant by a company, what is meant by a community within a company; what mutual responsibilities need to be recognized if that company is to be successful both in terms of creating maximum value added and in determining a just means of distributing that surplus.

One conclusion to be drawn is that whatever the strength in bargaining beyond the locus of an individual company, the continuation of the present extreme system is not only damaging to the health of the economy at large, but is not the best way of pursuing 'worker' interests within a company. The arguments about economic growth and the distribution of income, will continue to revolve around the ability of that establishment to create value added, the surplus between what it buys in and what it sells out. The prime interest of workers, as participants and beneficiaries in that industrial process, is in maximizing value added. To establish these conditions it is necessary to maximize the ability and willingness of people to work together and to create trust between them so that the task of managing is pursued fairly and effort is justly rewarded.

What would be the conditions for successful change in attitudes on both sides? Certainly more participation by workers in company leadership would help. Conditions need to be created in which the interests of all those whose lives are linked to the company are taken into account. However, whether or not to have workers represented on the operational board, must be judged entirely in terms of whether the change would improve company performance. Since the operational board members are, or should be, chosen for their managerial ability, it is not likely that a formal representative of the workers would help materially.

More participation in day-to-day management would also help. There are a number of areas where this could be improved substantially. One is in the provision of more information. Industrial activity suffers from the same desire for secrecy as pervades other parts of British society. The reasons why it is thought that relationships would be better and motivation maintained if information is held back from others are difficult to understand. It is also particularly irritating that so many companies excuse their refusal to give employees more operational information by hiding behind the insider information clauses of the stock exchange rules. Companies appear to be missing the point, that employees are insiders!

More open debate about both organizational and personal objectives and constraints would also be valuable. Communities and organizations are healthy to the extent that they allow the question 'why?' to be asked. It must therefore be true that the more management can open itself to questioning, from amongst its workforce about where it is going and the means it is using, the better will be the performance - even if the process becomes more time-consuming. Current management theory has done a lot of damage in this respect. Techniques for formulating and planning the execution of company objectives have reached a high level of complexity but it is often forgotten that the original inspiration (vision) of where to go and, at the point of implementation, the ability and the motivation to carry out plans, are human issues (people problems) not subjected to advanced mathematical techniques. Management techniques are of no avail if the company chairman has no vision or if, at the other end, those at the lowest level of operations end up by saying, 'Why on earth should we do that?'.

It is still a mystery why the concept of value added, is not regarded as central to the health of a company and is not used as the key yardstick, both of company performance and as the source of rewards. Both sides ignore it. Managers and shareholders are obsessed by profits, unions by wages. Yet the *distribution* of value added between these two legitimate claimants, is subsidiary to the act of creating value added in the first place - and to do that successfully, capital and labour must work in partnership not as rivals, even if, later, there may be an argument as to who gets what.

The foregoing may go somewhat against the grain of popular perceptions of management, but it needs to be said that these preceding conditions are essential and have to be met to ensure success; they should not be factors that workers have to claim from a reluctant management. They should be regarded as the manifestations of good management. What the workplace requires is not less leadership (replaced by more participation) but better leadership; because good leadership involves participation, good leadership leads to more information being given, good leadership attracts respect and loyalty.

It has been mentioned a number of times that the legitimate interests of the partners in a company have to be protected. Workers, via their unions, feel that in many cases these interests are not recognised fully in a market system and that continual vigilance is required to protect these interests. Since labour is a flexible cost, in contrast to a machine which is a permanent fixed cost, it is particularly vulnerable and needs protection. There is, of course, a corresponding insecurity felt by managers and by shareholders, where worker power is seen as a threat. Often the chemistry of these two positions leads to conflict and results in a wide ranging ideological debate at national level.

It has also been argued that the primary element in defusing this situation and creating successful *and* just forms of economic activity, is to make the individual company the focus of attention, highlighting the task of creating value added as the primary objective; that is the best route towards national prosperity. It is also the best course for pursuing the interests of participants in that company - especially the workers.

There is a need for much more open debate about the nature of economic activity - a debate which must fulfil three objectives. First, to recognize constraints. It clearly has to be argued and argued strongly that companies are not free to do what they want. In most cases they have to operate in a harsh, competitive international environment that dictates what they can charge and pay, and what they have to do to gain business. The more these harsh facts are spelled out and the options open to coping with them explained, the better. The frustration of managers who feel other employees do not recognize the extreme pressures under which they are obliged to cope with these issues is understandable.

Second, objectives, and means of achieving them within the company, need to be openly examined. One of the justifiable complaints of company workers, and a constant irritation to them, is that they feel that by the time they hear often unpalatable news, all other options are closed. From their perspective, management, by controlling the timing of news and action, seems to ensure that whatever debate takes place will be on their terms. Often this results in a damaging and negative situation. It sows bitter seeds of resentment for the future. It has also presumed that earlier, advanced discussion of the looming crisis would not have led to possible better solutions from the workers - or at least an offer to forgo, for example, wages to gain time in order to explore other, better solutions.

Third, the respective claims on a company's surplus should be debated much more fully. There are a variety of claimants for this surplus. There is the company itself (to assure its future via investment), the owners cum shareholders, the employees and the broader national interests, including customers. A better and more positive debate about the division of responsibility and the strength of claims on a company, will be a major step forward but western societies at present find it hard to conduct such a debate.

Notes

1. Hobsbawm, op. cit. p. 98
2. Mantoux, Paul (1947), *The Industrial Revolution in the Eighteenth Century*, Jonathan Cape, London. p. 375
3. Mantoux, Paul, ibid. p. 377
4. Mantoux, Paul, ibid. p. 382
5. Mantoux, Paul, ibid. p. 384
6. Mantoux, Paul, ibid. p. 386
7. Mantoux, Paul, ibid. p. 387
8. Mantoux, Paul, ibid. p. 389
9. Mantoux, Paul, ibid. p. 397
10. Mantoux, Paul, ibid. p. 419
11. Mantoux, Paul, ibid. p. 424
12. Pollard, Sidney (1965), *Genesis of Modern Management: Study of the Industrial Revolution in Great Britain. New ed.*, Gregg Revivals, Aldershot, p. 184
13. Pollard, Sidney, ibid. p. 54
14. Mantoux, Paul, op. cit. pp. 427-8
15. Pollard, Sidney, op. cit. pp. 254-5
16. Pollard, Sidney, op. cit. p. 270
17. Pollard, Sidney, op. cit. p. 272
18. McGregor, D. (1971), 'Theory X and Theory Y' in Pugh, D.S. (ed.), *Organization Theory*, Penguin Books, Harmondsworth
19. McGregor, D., ibid. pp. 305-306
20. McGregor, D., ibid. pp. 315-316
21. Wiener, Martin J. (1981), *English Culture and the Decline of the Industrial Spirit, 1850-1980*, Cambridge University Press, is an important exception.
22. This book lacks the space to enter into the vast debate over defining work and non-work and the importance of being paid for what we do.
23. This latter analysis could also be applied to, say, German industrial relations with the important difference that German trade unions take their 'management' obligations more seriously and recognize that conditions gained through wage bargaining require a commitment on their part to support the efficient running of companies. See chapter 12.

9 Economic and social models

It is common sense to assume that economic activity and broader social activity are inter-related. It is natural, too, to want to establish some form of structuring of that relationship - both to add to understanding and to provide the basis for judging what are good, and what are bad, features of that relationship.

Reference has already been made to economic and social models. The term, model, is commonly used in modern economics; but in a number of related ways. It can describe a single relationship between a number of variables. Expressed in mathematical terms, it usually contains one or more independent (explanatory) variables, deemed to represent cause, and a dependent variable in which changes are to be measured as the effect. The relationship is invariably formalized to concentrate on what are regarded as the essentials in the relationship and it will be constructed in order that, by use of statistical, regression, techniques, it can be fitted to actual data. This process provides estimates of the quantitive relationship between the independent and dependent variables. Such modelling proliferates in modern economies. Fisher's equation referred to in Chapters Three and Six is one example. Another is the relationship between income, price and spending on specific goods (a consumption function). Another is the relationship between changes in the exchange rate and volumes of imports and exports. The feature of this approach is that it contains a stylized expression of relationships, within the body of economics, expressed in a way that allows them to be fitted to actual data. There is a wider sense in which the term, model, is used. This refers to a mathematical formulation of a system combining such individual relationships, already described, to provide a comprehensive explanation (model) of the workings of the economy as a whole. The formulation of such a model and the study of its characteristics, together with its use to predict behaviour, comprises a large part of what is called macro-economics.

The economist's extensive use of models, particularly in these forms, is not common amongst other disciplines, but its main purpose is to seek to formalize basic relationships of cause and effect for individual variables, or for a system of equations as a whole, so that they may be fitted to data by statistical methods to obtain quantitive estimates of parameters.

The model thus represents a set of variables selected and related in a way as to help explain economic activity. The model must clearly indicate what phenomena it intends to address and which it proposes to ignore. The criteria for this choice are partly pragmatic, i.e. is a particular area of observation relevant to explaining parts of the model? Or they may be ideological, i.e. they are included or excluded by virtue of the fundamental values stance brought to the task. The task of model building requires the spelling out of hypotheses; relationships which link together cause and effect. They may be micro relationships, i.e. those that seek to explain how individual elements of the model are related - for example, the relationship between the price and the amount supplied and demanded of a particular product. They may be macro relationships, i.e. the way such individual equations are brought together and aggregated. Finally, although it is less often and less easily articulated, the model requires a value base; that instinctive ideological stance incorporating beliefs which determines what is regarded as interesting and relevant and what is not.

So much for the use of models in economics But considerable difficulties arise when an attempt is made to transfer these approaches to an analysis of social activity. There are no clear boundaries to work to. The term, social, embraces a vast range of activities that make up daily living. It is not easy to formulate and test precise relationships to provide quantitative estimates of the importance of explanatory factors. It is impossible, for example, to say with confidence that, if parents were to revert to using corporal punishment at home, there would be less juvenile crime or stronger nuclear families. Sociologists are a long way from achieving such precision in their analyses of social behaviour. It is also clear that many different value sets underpin society and it is hard to identify common characteristics.

What is common sense, however, is, to use the terminology of the model, that the social model embraces more than the economic model and, in most respects, the economic model has to be regarded as a subset of a wider social model. This point was made at the end of Chapter Six. Each subset of society may have its autonomy but it cannot operate without regard to others, especially the over-arching social model. The latter provides the ultimate framework. It has many parts of which the economic is only one. It sets the scene in which all else operates.

Returning to the economic model for a moment. The principal claim of the supporters of the market capitalist model is that it effectively delivers the main objectives of economic policy. It allows choice to be expressed and allocates resources efficiently. It, normally, produces more resources this year than last. It provides the means for achieving economic stability, such as low, sustained, inflation and balance of payments equilibrium. Finally, the economic model provides a rationale for distributing income. Western economists, businessmen and politicians including all but the extreme left wing, believe they have in the market capitalist model the best mechanism for achieving these objectives, provided certain conditions are met.

First, the concept of the market should be applied wherever, and whenever, possible. The market is not the only means by which people can exercise choice, but in most parts of economic activity there are few alternatives to match it. Provided people have purchasing power (an important proviso) and provided the product or the service is one which lends itself to market evaluation (an increasingly important proviso too) the market capitalist model provides the means by which a vast range of economic choices are made. It follows from this that society would do well to accept, within limits, whatever needs to be done to ensure that markets work as effectively as is possible. Competition policy, anti-monopoly legislation, consumer orientated activities - all have an important part to play.

Second, the firm and, in particular, the entrepreneur must be allowed free rein. At the heart of the operation of the firm is the risk-taking function, and the system depends heavily on giving ample scope to those taking such decisions. It also follows that for the system to work effectively the key players must be adequately rewarded for the part they play. This means adopting a relaxed view towards income distribution, letting the market decide who is worth what, never mind the many idiosyncrasies that may arise.

Third, many, but not all, supporters of the market capitalist model would argue that government has a small part to play in the detailed process of economic activity. Its role should be limited to providing macro economic stability and the appropriate legislative framework to preserve the maximum amount of freedom for economic agents and markets.

These statements about the objectives of economic policy, and about the workings of the model designed to meet them, have, however, their own social implications. Methods of remunerating those engaged in economic activity, for example, raise issues of both freedom and justice. The emphasis on freedom as a characteristic of underpinning market activity, and the motivation of the principal agents, also implies a particular view about responsibility. The ideological basis of the view about the appropriate role of government, may have wider social implications. In fact, the economic model, even as described in this summary form, creates a sort of social model of its own. Despite an intuitive view that the economic model is, and should be, a subset of a wider social model, and dependent on that wider social model, in practice the day to day impact of economic activity, associated with society's obsession with immediate material objectives, appears to constrain the form the social model should take.

The argument can be made another way by looking through the eyes of a hypothetical manager of a business. The manager of a business is there to maximize a surplus. A precise definition of that surplus will be deferred for the moment, for reasons that will be clear later; broadly speaking, it is the difference between sales and costs, between outputs and inputs. The size of that surplus will be determined by key economic and technological parameters; the quality of the product, the efficiency of production, and the product's place in the market. However, once the surplus has been established by those broad parameters, the manager has a number of internal or consequential choices to make, which represent demands on that surplus, some of which he may be comfortable with, others less so.

One such demand would be for funds for the training of staff. No manager, if he is sensible, would begrudge an adequate training budget. It is akin to spending on new plant and machinery. The latter improves the quality of the capital stock, the former improves the quality of labour as a factor of production. So far so good. Another request for funding (most of these questions will undoubtedly be presented by the human resource (personnel) manager), would be for generosity about workplace welfare and fringe factory benefits. Some of these, such as pension schemes, are virtually imposed by national and industrial good practices, but others, such as the quality of amenities like canteens, and sports fields, are much more open to argument. Much would depend on whether McGregor's theory X or theory Y style of management is being practised. In today's climate it is certainly not obvious that amenities of this sort are likely to be provided without question.

Another request which is potentially time-consuming and costly, which, again, would emanate from the unfortunate personnel manager, is for the manager to accept the role of union bargaining and encourage worker participation. A manager is inherently one whose is paid to act decisively; to identify a problem, find a solution and act to implement it. Often the traits in the character of people with these skills make them impatient of related issues such as the requirement to consult and to persuade others to follow his or her leadership. It is not surprising to find a natural resistance to union bargaining and worker participation, especially as union bargaining inevitably raises wider issues other than those strictly related to the individual company's performance. The manager would rather do without this element in managing the business. Of course, he or she cannot, but with eyes set on retaining as large as possible surplus, union bargaining and worker participation is not a welcome issue.

These claims on the surplus listed so far reflect a rapid descent down the manager's scale of approval of the agenda brought to him by the personnel manager. But there is more. After the personnel manager has left with the best deal he can get, the finance director may come in to go over the accounts and will quickly draw attention to the amounts allocated for general taxation. He will remind the manager that they are the consequence of politically agreed social policies designed to fund services, such as health, education, social welfare and the pursuit of redistribution policy. All good and worthy objectives maybe, says the manager, but, again, they detract from what he is really there to achieve.

What he reckons he is there to do is to maximize the surplus for two principal purposes. The first is to secure the company's future by ensuring that enough funds are available to maintain and improve the company's capital stock. The second is to satisfy the owners (shareholders) of the company by providing an adequate rate of return on their capital.

From the manager's point of view this succession of claims draining the surplus (primarily to be used in these two ways as he sees it), however worthy and necessary those other claims may be, are regarded as costs to the operation of the company. They limit the size of the disposable surplus and will, therefore, be resisted to a greater or lesser degree.

Moreover, returning to the problematic issue of the definition of the surplus, an even more important consideration faces the manager. If he takes the view, as many managers would, that the wage and salary bill is to be regarded as a similar cost to be charged against the disposable surplus (and argued over in dialogue with the unions, for example) then the task of keeping the wage bill as low as possible becomes one, and perhaps the key, task for the manager. The implementation of that task becomes adversarial and the remuneration of the employees is placed in a different category to the rewarding of the shareholders.

There is a different view. In strict economic terms, the surplus that is maximized is the difference between sales and purchases of the company, the latter being all items, goods and services, bought in from outside the company. The economic surplus is therefore the value added to those purchased goods and services, created by the combined efforts of labour and capital deployed by the company. The value added is created by a cooperative effort of all internal factors of production. The claims of employees on the surplus (added value) are as legitimate and in the same category as those claims to secure the company's future and those of the shareholders. Of course, the claims need to be addressed sensibly and there is room for immense debate, which will no doubt be heated, about the ultimate allocation of added value amongst those who have a legitimate claim on it. Nevertheless, the distinction being made is crucial. Value added is created by the cooperation of factors of production. All factors of production have a claim of the same intrinsic nature and value as each other. It confounds the principles of management and crucially destroys relationships, by erroneously treating the reward of one factor of production as a cost chargeable in calculating the surplus and to treat other claimants differently.

The main reason for using this little illustration is to make four intimately related points. The first is that from a manager's point of view, and it would be shared by many, workplace concessions and welfare are seen as costs on the enterprise. Many of them may be inevitable and some of them may even be welcome, in the sense that they assist in the pursuit of desirable objectives, such as training. Nevertheless, they are still regarded as costs and are therefore to be kept to a minimum.

Second, the whole nature of the calculus, and of the attitudes that are brought to bear, reflect the underlying assumption that the manager, since his task of wealth creation (creating value added) is paramount, must be allowed to define both the problem and the solution. That is why, when major redundancies are announced without prior consultation or without an open debate about the alternative solutions, the action is generally tolerated. It is assumed that management has the responsibility to create wealth and must be allowed to get on with the task, and do it in the way management should direct; another version of the Americans' tag 'business is business'. Moreover, it also leads to the adoption of management approaches which are heavily loaded in favour of this management culture, making sure that the responsibility for both problem and response are kept securely within management's hands.

Third, and, more generally, 'welfare' in these terms, is seen by management as a luxury where the ability of the business community to respond to these 'legitimate' social needs is dependent upon success. In good times business can afford to be more generous, in bad times business must batten down against the storm and cut down on luxury items such as welfare. This attitude not only ignores the positive contribution of those 'welfare' functions in ensuring the success of cooperative management styles. It determines that the debate about levels of welfare (and the business community will be as vocal about the need for 'cost' control of general social programmes as about their own internal welfare issues) is carried out in the context of what is permitted by the current economic climate.

Finally, it follows from these considerations that if wider matters of social policy are allowed to be approached in this way, the approach will inevitably be based on the widest possible scope for the role of business values in society. The climate will be set by the prior conditions emanating from a particular business view about what needs to be done (or avoided) in order to maximize the economic surplus.

Taking all these factors combined, it is clear that business is allowed to define many social parameters. The latter have to bow to the exigencies of economic activity. So much, therefore, for the a priori view that the economic is dependent on, and subservient to, a wider social reference.

As has been said already, it is not possible to define social activity precisely; its characteristics, its frontiers, its inter-relationships, reflecting cause and effect, or, most importantly, its value floor. It is a mixture of activities devoted to a variety of objectives which are not necessarily consistent with each other. It is, for example, difficult to combine policies designed to offer citizens all the individual freedoms they would wish for, with those designed to apply principles of justice. People cannot expect total privacy in their leisure activities in a society where the vast majority have the means to seek similar leisure activities. There is a finite limit to the number of private weekend cottages that can be built in isolated areas, for example.

The same is true about the means of achieving objectives. There is no reason to expect the many alternative means pursued in society necessarily to be consistent with each other. Current debates about the role of the market in social activity or new articulations of the role of the state, the role of consumer organizations or self-help groups, reflect debates about choice between alternative means of working to achieve social objectives. There is simply no reason to expect these alternatives to blend into a harmonious whole.

Finally, the social model is built on a variety of value bases which differ and are often inconsistent with each other. Much has been said about the choice of alternative value sets as a basis for economic activity, but the social perspective offers an immensely grander view of the amalgam of values to be found in different parts of society. Certain value sets are adopted in certain sections of social activity and they have important consequences as individual and institutional behaviour is developed around them. The values of compassion and care underpinning much of medical and community care is a prime example. Tolerance, or the lack of it, in a multiracial context is another. Besides the richness and depth that these individual elements bring to overall social well-being, there can be friction at those points where these differing value bases touch. Reality reflects a difficult and often ambiguous co-existence between these value sets especially when the frontiers are on the move.

Given these inherent inconsistencies in the pursuit of different goals, by different means based on different value sets, it is a hugely difficult task to articulate a social framework in the same sense that the economic model has been described. Optimizing a 'social model' by the use of policy tools, in the way that the issue might be addressed in an economic model, is just not on. Knowledge is too limited, understanding of desired social objectives is too blurred, the choice of instruments to effect change is unclear and there is a wide confusion over ideology; either in the debate about which, if any, of rival ideological claimants is superior, or in the debate as to how rival ideologies can co-exist.

Nevertheless, people are constantly seeking to achieve goals in their daily lives and a feature of western society is that many choices towards achieving those goals are made in the market place. The results are reflected in the massive spend on private consumption which dominates western economic activity. There is a social pattern of activity which uses purchases of goods and services obtained through the market place, to achieve specific goals of individual, family or other forms of human fulfilment. Moreover, the allocation of resources to meet society's goals in the areas governed by social choice takes place against a much more clearly articulated background of goals and methods. Since priorities in this area are decided through political debate this consequence is natural.

Many of those political, social, decisions relate to the provision of welfare and care. However, the achievement of goals in these areas is not limited to state activity. It is possible to talk of a welfare society rather than (more narrowly) of a welfare state. Many agents contribute to a welfare society. Personal action which, even if limited by circumstances, is still central to coming to terms with disadvantage. The family is still a principal caring agent. Many other agencies, the churches, charitable foundations, self-help groups and industry itself, provide welfare facilities. All these routes, to providing welfare, coexist alongside the state, as provider, and their activities need to be considered, if a full discussion of welfare objectives, and the routes by which they are to be pursued, is undertaken. One useful route for so doing is to develop the concept of a civil society.

Welfare policy today seeks to achieve a proper balance between rights and responsibilities. Social history is full of examples of the battle to achieve recognition of basic human rights. Many countries have such rights enshrined in a written constitution. There are international conventions, with recourse to international courts of appeal, where principles, involving the breach or protection of human rights, can be tested. The British position, of an unwritten constitution and parliamentary sovereignty, creates a difference more apparent than real. These historical and social forces have been successful in establishing fundamental rights, of liberty, for women, and for people of all races. An articulation of basic rights may provide the only moral basis for the fight against injustices. However, human beings have responsibilities as well as rights, and there is a moral case for arguing that as great an emphasis should be placed on the former as on the latter. There is, indeed, a religious argument that people have only responsibilities. In the context of a modern affluent western society, a case can certainly be made that people have become too concerned with rights rather than responsibilities.

The debate about rights versus responsibilities also raises a more practical issue of the cost of providing the right. Some rights may be virtually costless to provide; the right to life, to freedom of speech, and to equal sexual treatment. Others, and they are the ones in contention in the current debate about social policy, are not. Some may claim a right to housing, but who is prepared to accept the responsibility of bearing the cost? Have people a right to a job? To university education? In today's context, it is necessary to turn the question around. Who is willing to accept the responsibility of paying for the resources to meet that claim?

Just as the differing ideological bases of alternative economic models have been noted, the approach to social welfare policy is equally ideologically determined. It is often useful to identify alternative approaches to welfare in the following way (with examples to illustrate) [1]

Figure 9.1
Alternative approaches to welfare

Instrument	*Type of Policy*	
	Residual	Institutional
Selective	Bus passes for the elderly poor	Bus passes for all the elderly
Universal	Child allowances for lower income families	Child allowances Free health provision for all

Selective social policies relate to those where the objective of the policy is to assist a clearly defined sector of the community and for it to apply to no other. An example is the provision of concessionary bus fares for low income pensioners. Universal policies are designed to offer a welfare provision to all who fall in a certain category regardless of economic need. An example in Britain would be the provision of child allowances to all mothers with a child under the age of sixteen, regardless of family income.

The difference between residual and institutional welfare policy is different and is ideological. A residual welfare policy assumes that, provided people are allowed to get on with the task of working and earning a living and to accept their responsibilities, most welfare problems will be resolved (either by individual or family action). There will be some (residual) problems relating to the disadvantaged or perhaps the deserving poor, that remain the state's responsibility, but they will be minor (residual). The institutional view of welfare is both more pessimistic and more interventionist. It assumes that the modern industrial state is so complex that there will always be some, perhaps many, who are unfairly suffering from its workings. People may want to work but there is no opportunity; they may live in a disadvantaged neighbourhood; they may suffer from the consequences of anomie; or lack of close relatives, friends or other carers. The problems and resulting suffering, are alleged to be so numerous, and the arguments to intervene to safeguard basic human rights so pressing, that a modern state must offer an institutional solution.

Although this debate is often conducted at a technical level, debating what policies are best and most effective, at heart is a fundamental ideological debate that mirrors the debate over the choice of the economic model.

Those supporting a residual view of welfare usually argue that the institutional approach creates dependency. If people are encouraged to believe that hard times are not their fault, and that there is state apparatus ready to assist them, the incentive to improve their position by their own efforts will be seriously diminished. Such a dependency culture is an obstacle in the way of encouraging self-reliance and independence. The institutional approach attempts to deal with the problem of stigma - the separating out and the categorizing of a certain section of the community, in a way which may damage personality and hinder life fulfilment. Residual social policies make it easier to identify those receiving benefits. In the provision of school meals, for instance, institutional welfare supporters would argue that school meals should be provided to all children regardless of means. Otherwise a most unfortunate and damaging separation takes place at school based on the income of the parent.

Most of these foregoing issues are raised when the wider social model is approached from the point of view of the economic model. But it is time to return to look more closely at some other aspects of the wider social model in its own right. Our starting point is the presumption that, by beginning with the wider social framework, we will find the means of creating alternative links between the social and economic model which may be of benefit in assisting the economic model to work better.

An example from the area of business management will illustrate this point. One of the social changes of the past twenty to thirty years has been the public's, and in particular, the younger generation's attitude to authority. The change is to be found in the home where children's attitude to parental authority has significantly altered (maybe encouraged by a more liberal attitude by parents). It has spread into the school and affected the pupil/teacher relationship. In the wider community, similar changes affect the public's respect for the police and the administration of law. More widely, especially via the media, the degree of respect for institutions, for political leadership, for the churches, and for the state establishment, has diminished. The deferential society has disappeared, as some commentators have put it.

No judgement is intended as to whether this sea change in attitudes to authority has been good or bad, but it is useful to relate it to what has been happening or not happening in the area of business management. Since the second world war, the commonest form of management model applied in British industry has been based on military forms of leadership and organization. This is not surprising. The bulk of industrial leadership in this period came from the officer class with wartime experience. It explains the military terminology which has been so prevalent in industry. It also explains the enthusiasm with which industry embraces methods requiring individual adherence to detailed plans. These approaches so closely parallel the military model. Define objectives, propose plans designed to achieve those objectives, identify clear cut tasks required from all concerned. All that is then required is an unquestioning, dedicated and resolute response from everyone at every level of command.

What is supposed to happen when the view of authority implied by this management model, unquestioning response to undisputed leadership, is rejected? Rejected by the new generation of recruits educated and trained in a new social atmosphere where obedience to or sufferance of this type of leadership is no longer acceptable? Something has to give. Either the new recruits will be dragooned into this new and unfamiliar framework, and forced to work by uncongenial standards - a recipe for sullen unproductive attitudes - or the management model will need to be reformulated to provide alternative ways of getting the best out of people.

A useful way of generalizing this particular issue is by way of a Maslow-type pyramid (see overleaf). The following is not meant to be rigorous or exhaustive in its identification of issues, but simply illustrates a useful logic by which the argument may proceed.

The pyramid makes the following points. First, it categorizes social goals, ranging from the peak containing the highest statement of goals (the reader can choose his or her own specification) to the bottom of the pyramid where more specific and immediate goals are described. There are clear lines of descent contained in the pyramid. For example, one element in the good life is a sense of security. To have that sense of security requires a defence capability (against external aggression) and a system of law and order (for internal security). In turn that requires an army and a police force (as well as judges, courts and prisons). In any immediate period decisions have to be made as to whether the existing level of resources to fulfil these tasks are adequate or need to be changed.

A similar descending route is described for health as a principal element in the good life. It requires a health (illness?) service ending up with decisions on the suitability of present resources to meet present needs. The route to happiness for many of us, and in many areas, starts with a desire to acquire more consumer goods and services.

Lack of space in the pyramid prevents a comprehensive analysis but hopefully the point is well made. In addition, there are elements which ought to appear there to underline importance of this scheme of things. For instance, the activity of work is a contribution to happiness in its own right. It should therefore be shown somewhere in the centre of the pyramid. Below it should be included those things necessary to make work a more adequate contribution to happiness; training and education, security of job, variety of work, and good relations at work. The pyramid could have illustrated the preferred social processes and institutions by which society takes communal decisions. The national political process matters, but there are many other social community activities, where the form and quality of participation will have a significant bearing on the perception of the quality of life. A key ethical dimension in the quality of social life, is the means and extent to which society organizes its communication, the way in which society explores its value sets and organizes action consistent with them. The content of the pyramid is also determined by the choice of value sets (ideologies) since they

Figure 9.2 Social goals and means

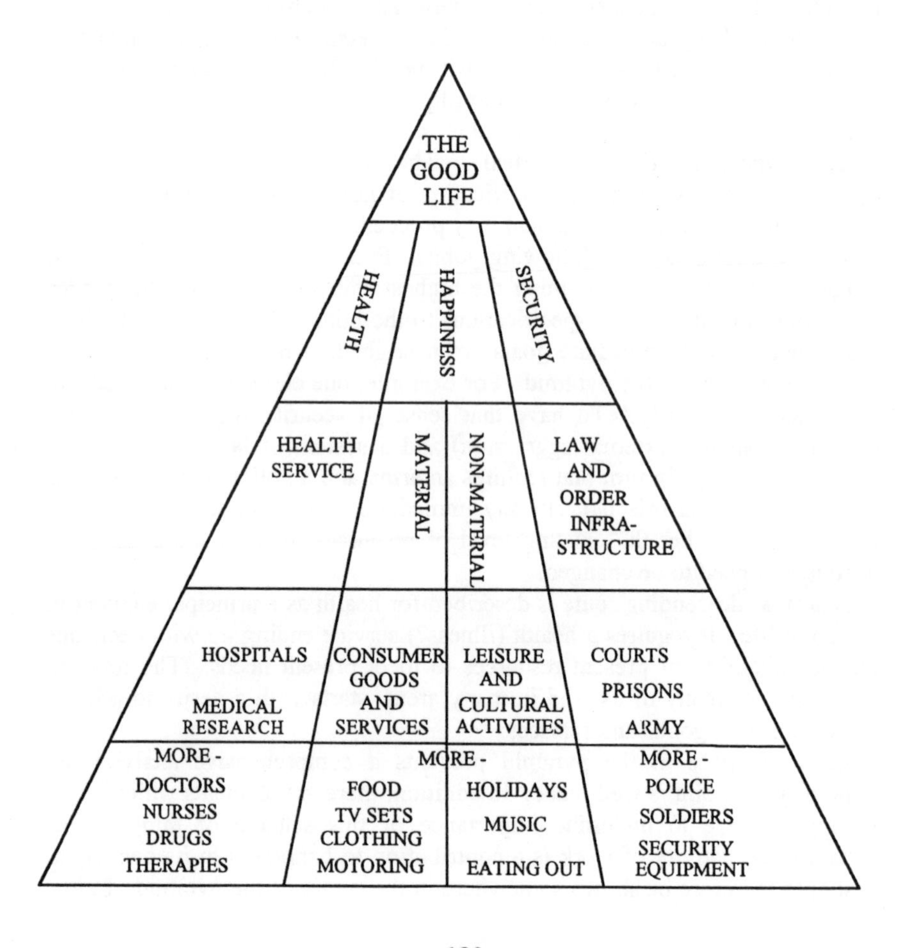

help to identify the balance of activities and the choice of routes to achieve goals.

The first conclusion to be drawn from this illustration is that the social framework sets the context within which issues of life fulfilment can be addressed. In that respect it stamps itself as superior, more dominant and more widely embracing, to the economic model. The pyramid also helps to make the point that, in the wider social context, non-material aims and objectives are supreme. Society's preoccupation with material things, goods and services, and the perception they matter most, derives from the fact that most political debate revolves around the bottom tier of the pyramid. The day to day debate is about the use to which this year's addition to the total of goods and services, provided by economic activity, will be put. The pyramid also contains many instances where cultural attitudes shape both the ends to which the fruits of productive activity are put, and the optimum means of delivering them. This point would be effectively reinforced if similar pyramids were constructed for other cultures.

Ultimately people want to be guided by this wider social framework. It provides the values against which to judge progress towards society's goals. It also provides the criteria (which have a value base) which indicate the ways of getting the best out of people in economic activity. The basic value sets in society need to be identified in order to determine society's preference for a particular type of economic model, what types of institutional frameworks it wishes to work within to achieve its objectives. More narrowly, to determine what type of management model it considers will get the best out of people.

The historical context of earlier chapters has, time and time, again illustrated the inter-relationship between the ideologies underpinning economic activity and those underpinning society at large, as well as underlining the changing nature of these relationships. Some would see, in this, a historical progression in which one system has irreversibly replaced another - a form of progress culminating to date in western capitalism and its associated libertarian ideology, consigning all else that has gone before to history. But that view, even conceding it has some validity as an interpretation of British history, would ignore the fact that in today's predominantly market capitalist world, other value sets (ideologies) clearly form the basis for equally successful economic models. The importance of culture and the relevance of the wider social context to economic activity should be capable of being tested further by comparing the attitudes to, and the form of economic activity preferred, in other parts of the world. This is the next task.

Note

1. Readers who wish to pursue these distinctions in more detail would find it useful to refer to Pinker, R. (1971), *Social Theory and Social Policy*, Heinemann, London, pp. 97-104

10 Economics and culture

Market capitalist economies are often presented as components of a universal system, with major international institutions (the IMF, the World Bank, GATT, etc.) created to monitor and police it. However, it is clear that the principal participants, Japan, the USA and Europe, enjoy immensely different cultural backgrounds. Since these backgrounds contain different ideologies they provide a useful opportunity to test whether the way they carry on economic activity is, or is not, affected by such differences.

An attempt to make cross-cultural comparisons is a daunting task; what follows will be necessarily superficial. There are also the dangers of generalization. Even in the familiar territory of British history, with which most readers will claim familiarity, there are deep ambiguities. The fact that other cases to be examined are based on cultures radically different from Britain's, and located many thousands of miles away, must warn against facile distinctions and judgements.

Another problem is to find a way of handling, safely and surely, the sense of historical perspective. That perspective may have a short or long time scale. In what period, for example, in talking about Germanness, should the argument be rooted? In an early view of the German tribes, who expelled the Romans? Or is it wiser to concentrate, say, on the era of Bismarck as the period which defines what is meant by German? In Britain's case, how shall the brash nationalism of the first Elizabethan era be reconciled with the global perspective of the Victorian period or the post-imperial adjustment? Which France is the more important? That with its inheritance in the early Roman empire or the post-Napoleonic era? Should more timely importance be given - in the case of all the principal players - to the post-second world war political situation?

Another dangerous course is to make generalizations, by eulogizing nationals traits selected out of many. Bismarck was a massive historical figure, as was Napoleon. Europe as a whole owes much to them in terms of the way in which they influenced the modern development of Europe. But the imperialistic nationalism that dominated the following 150 years, and led to so many major wars, as well as gross colonial exploitation, provides a dark side to European culture which must not be forgotten.

Thus, in pursuing the task of examining the importance of culture, the expectation will not lie in finding clear cut solutions or explanations, but, hopefully, in identifying reference points upon which to base some tentative conclusions.

Much has been written over the past twenty or so years, seeking to identify the causes of Japanese economic success - which in itself is a reflection of the extent of that success. [1] Most of that literature reveals the complicated nature of the task of understanding. Ambiguities relating to religious attitudes, the basis of personal relationships and the use of language itself, build up to a mysterious culture in the literal sense that is hard to describe clearly - even by Japanese writers.

In the longer term time horizon, Japan has always had to live with a nature loaded against its people. Despite the size of the islands, the amount of cultivable land is relatively small. The country suffers from a chronic lack of raw materials of the quantities required to sustain a modern industrial economy. The country is plagued with many earthquakes, large and small, and it is exposed to extreme weather conditions. It is no small wonder that the concern to survive ranks high in Japanese priorities. The concern is to be found deep in Japanese attitudes to life - an insecurity about what the future holds (this fear probably accounts for the very high savings rate amongst Japanese). It also accounts for the emphasis Japan puts on export performance needed to pay for necessary imports. It is interesting that in the oil crisis of the 1970s, Japan, entirely dependent on imported oil, responded so quickly to the massive adverse shift in the terms of trade that it restored its trade balance within eighteen months.

The underlying sense of a personal struggle for survival appears to give Japan a heightened sense of national purpose. From time to time this national view has been militaristic; culminating in the attempt to create an economic hegemony to secure its own material supplies that led to Japan's involvement in the second world war. Since then, it is as if the sword has been beaten, if not into ploughshares, at least into cars and computers. Economic growth and success appear now to be the focus of the need for a sense of national purpose.

Much has also been written about the nature of community in Japan. Whilst it is clear that the individual Japanese appears more than willing to subject his wishes to those of the group, it is by no means obvious why this is so. It has something to do with the need to gain respect in Japanese society, the respect for age and seniority. It also has something to do with the over-riding influence of, at least, group purpose, if not necessarily national purpose.

Whatever the deeper origins of the attitudes that the Japanese bring to bear on economic activity, the forms they take are easier to describe. They amount to a formidable and coherent approach to the matter of directing and managing a modern economy. The framework is partly institutional and partly composed of practices operating within each company.

At an institutional level the Japanese operate within a unique government framework for economic activity. Its financial sector also has a particular role to play. The way in which industry itself is organized is also largely different to the structures to be found in other western capitalist countries.

At government level, the most important features are the attention given to formulating strategic national economic goals and the, largely, indirect control of the national economy applied to ensure that those goals are clearly pursued by industry. The responsibility for achieving these objectives rests with the relevant government department, the Ministry of International Trade and Industry (MITI). It is hard to find a counterpart even in, for example, western Europe, where, over the past thirty years, that region's experience of economic planning would have created an expectation of finding similarities. MITI has virtually no legislative powers with which to enforce its decisions. Given the reality of Japanese methods of arriving at consensus, it may not even be true that MITI takes decisions in the sense Europeans or Americans would understand the term. But MITI's role in guiding Japanese industry towards a viable future, is undisputed and pervasive. Not all its views are accepted (apparently MITI resisted the move into automobile production in the 1950s), but MITI staff will be central to debate about exploiting new economic opportunities. MITI has been rightly named the anxious mother of Japanese industry.

There are other important characteristics of the system for which MITI is the hub. The business community willingly accepts the role of MITI in the knowledge that MITI is working in its interests - in marked contrast to business attitudes towards civil servants, with similar responsibilities in, say, Britain. Respect comes partly from the prestige gained by MITI and particularly in recognition of the intellectual powers of MITI staff. It comes also from the extensive interchange of staff between MITI and industry (again, perceived to be a recipe for bribery and corruption in western eyes). Finally, although MITI has few legislative powers, it exerts its immense influence, through having strong say in the provision of finance.

The banking system also plays a quite different role in Japan compared to that in other countries. There are parallels in the form of industrial banking that prevailed in the United States, until it was eventually ruled out as a restraint of trade, and in the more amenable German system. Japanese banks operate in a quite distinctive way. In the first instance they have a much greater financial involvement in Japanese industry. The loan/equity gearing ratio in Japanese companies is far greater than in the USA. As a result of this dependence on banking, rather than equity, finance, companies are able to concentrate on long term planning without having to pay too much (excessive in Britain and the USA) attention to current market judgements about performance. An obsession about quarterly or half-yearly performance as an indicator of how owners or potential owners of equity are likely to react, the common lot of British or American business, is rarely found in Japanese industry.

The corollary of this is that Japanese banks are far more pro-active in their relationships with the companies they have invested in than in Europe or America. They are partners in risk-taking and make themselves party to those decisions. They will place executives in key posts in the company; especially if, and when, a company gets itself into difficulty. The banking response is an injection of talent to help sort out the problem rather than a hasty withdrawal of finance to secure its own position. The transfer has a two-way benefit; it helps the company by providing additional talent, but it also gives banking staff practical experience of running a company.

The third institutional area, which again has no equivalent in Europe or the USA, where monopolies and anti-trust legislation is largely designed to prevent it, is the role of industrial associations. Japanese industry is highly competitive, between Japanese firms as well as with overseas companies. However, the over-arching concern of the national interest promotes a cooperative spirit between companies hardly to be found elsewhere in the world. Cooperation is informal, and difficult to pin down, either the process itself or the effective results of the process, but it is linked with the role that heads of industrial companies see for themselves. American executives complain of the waste of time attending meetings in Washington. Tea drinking, the Japanese equivalent of a Washington lobby, is willingly accepted as a role for their Japanese counterparts, with formidable results.

Within the company structure, the significant feature is the existence of families of companies, frequently with one large company at the centre of the family. Sometimes it is a commercial company such as Mitsubishi, and sometimes it is a bank such as Fuji. A feature is the way companies work together, particularly in tackling new markets abroad. Companies cooperate in strategic studies, even though, once the markets are identified, they compete with each other. Another feature is the way in which companies help each other out temporarily when in trouble. If a company in the family is running into problems, companies elsewhere in the family, that buy from it, will speed up their payments and companies elsewhere in the family, which sell to it, will delay their demands for money; this flexible position can make all the difference to the chances of survival for an individual company.

The internal practices of Japanese companies are equally distinctive. Much has been written about these practices, largely to relate them to Japanese cultural strengths. Some commentators attribute the adoption of a communitarian approach to business activity to the extreme devastation existing at the end of the second world war which created a need for companies to accept a wider responsibility for the fate of the local community. The good practices, which are so greatly admired by Japan's competitors, relate only to the major companies and, within those companies, to certain employees only. There exist many back street companies who are forced by the power of the larger buying company to cut costs to the bone, and treat employees harshly. Within the company, women, once they are married, do not share the privileges of their male co-workers. Also, the retirement age in most companies is particularly low with relatively poor pension provisions. That said, the management styles of major Japanese companies contain substantial advantages over their competitors. These advantages relate to the style of marketing strategy, the form of manufacturing methods and the procedures designed to enhance motivation.

An important element in Japanese commercial strategy is the approach to research and development and marketing. It was said that, at one time, Japan invested very little in research and development. Like Francis Bacon's comment on the English in the eighteenth century, 'they invent nothing and copy everything'. The first stage of Japanese economic development largely took this form, Japanese companies taking the view that it would be better to buy cheaper second generation technology, or make imitations, and rely on price competitiveness to achieve success. There are many examples of the pioneering efforts along these lines; buying, for example, an Italian Vespa in order to break it down in order to imitate, which led to the rise of the Japanese motor scooter and cycle industry. Success, at this stage, generated surplus funds, which has allowed Japan to build up its own research and development. Now it has the largest civil research and development programme in the world.

Japanese marketing strategy has many distinctive characteristics, the main feature being the degree of long term planning that precedes entry into new markets. Planning revolves around establishing desired market penetration. Once such targets are established companies rarely deviate from them. If difficulties arise, and targets are not being achieved, quite dramatic changes in, for example, pricing policy, will be adopted in order to get back to the desired market share target. Amongst other conclusions to draw from this strategy, is that once Japan emerges as a new competitor in a particular market, such is their pre-planning, the response can only be defensive and reactive. The chances of snatching the initiative back are small. It is better to spend the time anticipating where Japanese companies are next intending to jump.

Much has been written about Japanese production methods. The advantage does not appear to be attributed to more advanced machinery, although increasingly Japanese companies appear to be ready to be innovative in this respect. The introduction and use of robotics in production processes, is one such area. The adoption of advanced technologies, such as anti-skid devices in cars, ahead of competitors is another. However, the great advantage that the Japanese have over other manufacturers rests on their skills at organizing production. Their principal advantage has been in controlling stocks using the 'just in time' approach. This rigorous approach keeps down costs, cuts out unnecessary holdings of materials, of partly manufactured goods and of finished goods and has revolutionized manufacturing methods and in ways that are now widely adopted throughout the western world. Car manufacturing companies will now hold (and own) no more than a few hours supply of materials, the burden for ensuring these supplies are sufficient, and topped up in time, falls on the supplier. A complementary feature of this system is the way in which production flows are controlled throughout the system from the demand requirements to finished products, rather than the traditional approach seeing the flow running from raw material to finished product.

Although 'just in time' is a key aspect in Japanese manufacturing, its use also influences attitudes generally to improving the production process which reaps many cost cutting benefits by shaving minutes or seconds off production time, or by eliminating interruptions in the production flow. It also creates an atmosphere designed to encourage maximum participation by employees.

There are many strong points about the methods used to create motivation within Japanese companies. For example, Japanese workers are encouraged, and willingly accept the challenge, to be involved in the running of the factory; this relies on the use of quality circles, group management activities and by an open invitation to suggest improvements (with suitable rewards). It is said that Japanese suggestion boxes are always full. The main point is that Japanese workers maintain their own pressure to improve performance constantly. Japanese workers have a strong commitment to their own company which is reflected in the prevalence of company rather than industrial unions. A further advantage to the Japanese worker is access to information about his company's performance. It is said that Japanese employees will take accountancy classes simply so that they will be better able to understand their own company's accounts.

These characteristics combine to give both employees and companies a strong sense of company loyalty. Employees, in the larger Japanese companies, have a pride of belonging and a loyalty to the group within which they are set, whether it is a small quality circle, a marketing team or the company itself. Employees will often remain at their desks at the end of the day until a colleague has resolved a particular difficulty. In return, companies reward (at least a privileged majority) such loyalty by lifelong employment, wage levels that reflect family circumstances as much as productivity and a confidence in workers reflected in their increasing value to the company as their experience grows. Akio Morita makes this point in discussing what a Japanese company will do if an employee has made a mistake. [2] It is accepted that he is not alone responsible for the mistake. The group of which he is a member will be challenged as to what they did, or did not do, to allow him to make the mistake. The absurdity of sacking the employee is recognized, because, by so doing, many years of experience and training would be thrown away over just one blemish. Mistakes are also seen as a key form of learning and the expectation would be that the employee would perform better in the future.

Even this system is under strain if a company is in trouble. Something painful has to be done. It may result in employees being transferred to another part of the family of companies, involving moving many hundreds of miles away, or it may involve taking a severe cut in salary. However, whichever process is adopted, the Japanese employee can rest assured that the first to suffer will be the head of the company.

Whether other countries can imitate the strengths of the Japanese system or not, Japan is an example of how cultural strengths have been harnessed to the methods adopted for pursing economic success. Japan has built on community values and the power of the group (both ideological strengths deriving from social values at large) to motivate people at all levels and in all aspects of economic activity. It has produced a level of motivation envied throughout the world.

Japan also provides a modern example of the need for a country which embarks upon a course designed to catch up with other countries already well established economically, to do so with a prudent and cautious commitment to free trade. Like Germany in the nineteenth century, following the advice of Frederich List, Japan's main concern over the past forty years has been to build up its productive power and not solely, or necessarily, to maximize the gain from international trade. Only in recent years, as Japan begins to feel more confident in its ability, has it showed signs of wishing to be a full player in the competitive game of international trade.

The United States of America presents an equally interesting case. Looking at the roots of American society, an important indigenous factor, an influence that came from America itself as distinct from the cultural inheritance brought by immigrants, was the influence of the early pioneering spirit. A reckless attitude to life is not intended but rather the reality that, for a long period in the country's history of settlement, many Americans had to come to terms with open frontiers; the challenge of breaking virgin ground, both literally and in building new communities. Much has been written about the cultural shapes drawn by this experience, but it appears to have had one fundamental paradoxical effect. The early settlers developed a high degree of self-reliance. When a pioneer moved west into virgin territory, to put down new roots, he was taking on a challenge between himself and a hostile nature (that hostility including the reaction of people already there). Survival rested in his own hands. If he succeeded, he could be justifiably proud of the way his individual strengths had not been found wanting. The sense of personal achievement, resulting from self-reliance, goes deep into the psychological makeup of modern Americans.

The story of the early history of America also tells a different, complementary, tale. Alongside the personal struggle against the odds laid by nature, there emerged a basic sense of neighbourliness based on the realization that neighbours, who were few and far between in any case, were going through a similar experience. That experience included events to which an individual personal response was inadequate; the loss of buildings through storms, the need to extend buildings, illness, protection against robbers and thieves and the task of creating townships with schools and churches. Out of these needs emerged a strong sense of mutual support; willingness to help others in adversity and to celebrate, with others, the joys of practical community building. The warm tradition by which Americans still welcome the guest into their homes probably has its origins in these early days.

The early pioneers lived through a period which often tested their personal courage and resilience dramatically, yet at the same time gave them the opportunity to build communities, often before the arrival of effective law and government. Here is a key trait determining the American attitude to government. The basic attitude is not entirely anti-government but contains the view that government is not necessarily needed to achieve personal or social progress. The primary experience may not have been the principal determinant of the constitutional, political development that followed, but it is consistent with the traditional Jeffersonian view of government that the least there is of it the better.

There is another element about the pioneering spirit, which has had an influence particularly on American social welfare policy, or the lack of it. Because open frontiers always beckoned, succeeding waves of immigrants were always able to try somewhere else, if their first efforts to settle and create wealth were unsuccessful. In this fundamental sense, Americans do not need to look to others, e.g. the government, to help them out of their own troubles - they can always move on. In contrast, European nations found, in the nineteenth and early twentieth centuries, that there was no alternative other than to deal with, say, the social consequences of the industrial revolution, by creating social welfare policies and constraining the excesses of market capitalism. Up to now, Americans have always had another option - to pack up and move on.

These factors have contrived to produce a powerful American cultural base. It embraces the perception that society needs to be free to allow individuals to take care of themselves in a land whose natural bounty will always provide for them. It also looks to the family as the prime unit of care. The role of the church is to set moral standards. Government activities are limited to the basics of national security, law and protection and, to ensuring the smooth functioning of the working of the economy. Also there is a wide acceptance that business should be given free rein to create wealth and deliver the resources that Americans wish to have.

It has been argued, in earlier chapters, that the American phrase, business is business, is ambiguous if it is interpreted as meaning that business activity has, or requires, no moral content. Many businessmen try not to be challenged in moral or ethical terms. Nevertheless, business has a massive influence, often ethical, on American society hardly to be found anywhere else in the world, not even in Japan. The activity of the business sector is seen as crucial to social development, especially the pursuit of an ever higher standard of living. It is also regarded as the arena for the practice of a modern puritanism, as demanding of individual, as anything to be found in the Victorian era in Britain. The morality of working and of succeeding, the acceptance of personal responsibility for failure, the morality of the market place itself, are essential ingredients in the perception Americans have of the way to seek self-fulfilment, earn respect and judge satisfaction and happiness. The moral dimension is positive, not narrow or negative, and, in many respects, exciting. It can be seen to be practised in many ways from the active interest taken by stockholders in their corporations, and the prevalence of corporate social responsibility statements.

Whatever the theoretical and philosophical strengths of this view of American culture, the American's view of life and his or her rationalization of his or her behaviour, both personal and social, is pervaded by the fact of success. Americans claim that their system works. It has produced the highest per capita levels of income achieved so far anywhere in the world. It has produced a massive corporate sector containing many multinational companies of formidable international repute. It continues to operate from a resource base of land and people which is the envy of the world. It represents (or likes to think it does) the ideal model of market capitalism and, through its vast network of business schools, pours out the high priests of this faith to proselytize the world.

An integral part of this economy is a formidable industrial structure. Corporations are largely driven by stockholders' perceptions of immediate performance prospects. There is a heritage of industrial banking supportive of business. The corporate sector supports its own research and development activity through corporately financed research institutions, such as the Rand Corporation. If the research effort of government, through its defence and space activities, is added in, the result is an impressive effort ensuring an adequate flow of new technologies for future generations. American business is also noted for its marketing emphasis. The salesmanship qualities of

American business are both pervasive and persuasive. America has the advantage of a huge internal market, within which to hone its products and practise the art of business. At present both imports and exports represent a small proportion of the total market. Such relative immunity to external world conditions provided by an enormous home market, have given America a considerable competitive edge over other countries.

Despite protestations to the contrary, the United States has a very effective planning system in place. It is ideologically incorrect to talk about planning, in the sense of state planning, in the context of the United States. But, standing back from the ideology and addressing the more general question whether the United States has mechanisms for taking a view about the future it wants and how it aims to get there, the answer to this less politically offensive question, must be in the affirmative. Through corporate research and development activity and through the state role in the defence sector, there is an enormous concentration of effort looking at the shape of the future. The activity is effectively reinforced by the work of many free ranging academics such as Kahn, Toffler and Naisbitt. Moreover, in less exciting, although more politically sensitive areas, such as agriculture, the United States has never been slow to intervene and regulate when necessary. It has also built in many institutional features which control, regulate and direct activities, such as the Securities Exchange Commission and the anti-trust bodies. Also, the interventionist new deal era of Roosevelt and the new society approach of Johnson are legitimate parts of American history. All this activity, whilst not normally grouped together systematically, gives the lie to the United States as a freewheeling, undirected economy.

The American influence on the international economic system has been substantial and not least in respect of the new discipline of management studies. The volume of academic writing in this field is formidable. So, too, is the influence of management gurus, each with his or her particular personal style of goading business into new styles of managing. The world has looked to the United States for most of the modern techniques. What the world, particularly in Britain, has not been so quick to do is to recognize that these techniques, although presented in the guise of the discipline of management and thus value neutral, are as culturally determined as any other element in economics. The British can be forgiven for falling into this trap. After all, American text books on management are written in English encouraging the assumption that the message they contain is equally easily transferred into a British context.

That assumption misses the point; these management techniques reflect why Americans are successful but they do not necessarily describe what others with other cultures have to do to be equally successful. American management models revolve around structures in which there is a clear identification of individual responsibility, where there is a provision for individual reward and individual punishment, and where financial reward is regarded as the main incentive. If to this list is added the American instinctive ability to respond and cope with challenge - the fix it mentality - the resulting list of qualities that underpin American management is formidable indeed. The list is, however, culturally derived and unique to Americans' perceptions of how to organize men and women, to elicit the maximum motivation in the community called business.

The American model is distinctive both in origin and in practice. It may be that Americans, by their success in the area of business and economic activity, have a claim to having discovered universal truths which all countries must pursue if they aspire to the same success. However, it is more likely that their success stems from the American ability to harness particular cultural strengths into the business model.

What is surprising about the United States is that it has not replicated many of the institutional reactions to the adverse effects of industrial change commonly found in Europe. It is not possible to talk about social welfare models in America in the way they are debated in the European context; this is strange. Industrial development in America produced the same tensions in terms of harsh working conditions, excessive hours, abuse of child labour and inadequate safety. It produced the same large industrial conurbations with the same anomie. It created a rhythm of economic activity which, from time to time, created massive unemployment which, by no stretch of the imagination, could be attributed to the laziness of the individual worker. Yet these conditions did not lead to the consideration of comprehensive, state-based, social legislation to deal with the four evils of poverty, disease, unemployment, and lack of education, as Beveridge described them.

The emergence of the political dialectic of left and right in Europe, which can be attributed to the same industrial development, has not occurred in the United States. The division between Republican and Democrat does not carry the same distinction as, for example, between Labour and Conservative in Britain. In America there is an uncanny absence of an ideological debate which is taken for granted in Europe. Nor does American society seem able to admit to the term 'working class'. Class is a European and not American concept, yet a large part of the American labour force works in jobs and relates in power and management terms in the same way as industrial workers in Europe. Why is this distinction not reflected in some form of, de facto, class structure as in Europe?

The cultural elements already described may explain the difference between America and Europe. The individualistic attitude - allegedly held by the vast majority of Americans whatever their status - towards earning a living is unique in its intensity to the United States. American workers - certainly American managers - wish to be judged on their own personal performance. If they are successful they will be impatient for reward. If reward is too long coming from their existing employer they will use their success to find a better paid job with someone else. If they are failing, they expect to be fired. The speed of action and reaction, of hiring and firing, is mirrored in the speed with which corporations seek to get back invested funds. It reflects a virility in American business activity and is cultural in origin. It is a product of a culture that expects its children to set up stall in the back yard to sell lemonade, or newspapers at an age before European children are even let out of their parents' sight.

The American cultural pattern is impressive, but, as in the case of Japan, it has to be asked whether all is as it seems. For all the argument in favour of market competition, both its ideological and practical advantages, America remains an example of shrewd planning and intervention done in the name of national interest. The willingness to act in that name is never far below the surface. There is the inclination to protection in international trade, based on domestic interests, which is no less, even if no greater, than in other countries. What has impressed, up to now, about the American story, is not any theoretical articulation of the reasons why, but simply the fact of success. Success is all important - it justifies all.

However, success raises a crucial question for the United States. Like Britain in the mid-nineteenth century, the United States has been enjoying a favourable conjunction between its cultural values, the indicators of personal success, the forms of esteem and the reward looked for, and business values, what is expected of individuals in corporations. It is not surprising that, at present, the United States is preaching adherence to free trade principles everywhere. From its position of strength, again like Britain in the nineteenth century, it is in its interest to do so, especially now in the field of services. But Americans are feeling the pressure of competition. Other countries are now overtaking them in terms of per capita income. Its own growth has remained near to one per cent per annum for many decades. The fact of slowing down, and of relative decline, is now a reality. If the image of invulnerability is challenged and the underpinning justification, success, is removed, what will be left? No major economic power, after rising and maturing, such as Spain in the sixteenth century, and Britain in the nineteenth century, has yet discovered the way to effect the massive change in cultural attitudes to generate economic rebirth. The United States will have a particularly acute problem coping, in view of the central role it has given to business culture in society.

It is a moot point whether it is legitimate to offer an analysis of the European region in the same way as Japan and America; as if the region is a homogeneous unit. Modern history spanning the past five centuries, records a past largely revolving around the emergence of European nation states, that gave scope for the expression of particular values and characteristics of community in the nation. That past also records the consequences, as nations battled to defend their own immediate interests and to acquire trading interests throughout the whole world.

How permanent are European characteristics? There are many Europes. There is the Europe of the early Roman empire. There is the Europe that followed the empire's dissolution; the resultant ethnic mix as different peoples overran and settled in different parts of the continent. Later, there is the Europe of the Holy Roman Empire, which defined in many respects the concept of Europe, a common culture, building on a rediscovered classical civilization protected by an over-arching political power. Then there is the Europe of the nation state, which encouraged particular interpretations of Britishness, Frenchness and Germanness. But the Europe of the post-second world war period is different as these many parameters relating to Europe's future were reset, especially those on which the present drive towards unity is based.

If a start is made by looking at Britain (although in considering cultural characteristics, would it not be wiser to talk about English, Welsh, Scottish?), the country's history contains many elements that rest intensely on a sense of national identity. The works of Shakespeare are full of such nuances, 'This sceptred isle, ... This happy breed of men, ... This precious stone set in the silver sea'. [3] Written in the full flood of national consciousness of the early seventeenth century, this insular emphasis was political in the first instance but cultural as well.

Many of Britain's traditions have developed differently to those of its neighbours. The concept of the rule of law is one such area. It enshrines in common practice much of the common sense of what peace and order mean in a British context. Sir Ivor Jennings [4] in his book, *The Queen's Government*, starts by recording a conversation between a policeman saying, 'Hey, you can't do that 'ere' and the instant Cockney retort, 'Ho! can't I, and 'oo says so?'. This exchange aptly sums up much upon which the country's legal system is based in contrast to various legal codes to be found in, say, France or Germany. Britain also owes much to the fact that it has had practically no period in which serfdom has prevailed. The democratic roots were put down as early as the eleventh century.

Englishness has also been allowed to persist, perhaps more strongly than otherwise, by bequeathing to the world the nearest approximation to an international language since Roman times. So much of the world uses English as the lingua franca that this has allowed British people an excuse for not having to work too hard to relate to others.

Besides these indigenous elements of Britishness, Britons have been able to avoid facing up to their European identity, such as it is, by two other major historical facts. The first is the history of the empire. For more than a century Britain considered itself self-sufficient, at the heart of a vast worldwide empire. The relationships with the empire fully satisfied its sense of identity, politically, culturally and economically. Notwithstanding the fast decline of the empire, echoes of the imperial past have been sufficiently strong, particularly in the creation of its latter day version, the British commonwealth, to continue to appear to offer, up to very recently, a real alternative to European integration.

The second major historical fact has been the link with the United States of America. A major proportion of early settlers in America were from Britain, which ruled most of the region until independence, English was adopted as the national language and Britain's economic involvement in the region's development, particularly during the nineteenth century, was massive. These factors forged links that have remained in place, even though the balance of power has now swung the other way. Many Britons look favourably at continuing links with the United States, even to the point of being content to become the fifty second state of the Union. From their point of view, Americans feel very comfortable with the link, particularly because of language and history and more, self-interestedly, because of the usefulness of Britain as a base for working with the rest of Europe.

There are many ways in which Britain, for good or not so good reasons, can be tempted to avoid the issue of its European identity. Yet its history has been as intimately European as any other nation state on the continent. It shares common political traditions both on the right and on the left. It shares a common political history. Its industrial and political experience, since the beginning of the nineteenth century, has resulted in the creation of common welfare approaches. Culturally, along with at least France and Germany, it has shared a sceptical view of the advantages of unbridled capitalism - a trend condemned equally in catholic and protestant circles in the major countries of Europe. Welfare principles are largely common, as is the approach to industrial relations.

In contrast to Britain, instinct, in examining similar issues in Germany, is to think in terms of a much shorter time scale; that instinct is sound. Germany's modern territorial boundaries have only existed for some 100-150 years (indeed, even this modern Germany has suffered the trauma of division for some fifty years since 1945). Despite this sense of historical newness there is a legitimate and wider historical canvas. In the widest sense of being German (the context of those speaking German) Germany has had, perhaps, the longest and most pervasive influence of all in, especially, central Europe. There are commonly perceived national habits, such as the sense or order and discipline and a particular acceptance of the role of the state.

Three important historical periods do have to be considered to make sense of Germany's current attitudes towards its own society and towards Europe. The first is the crucial defining period under the leadership of Bismarck. The second and third relate to the period since 1939. The first of these is the cataclysmic impact on Germany's relationship with its European partners, resulting from the Hitler regime, and the need to avoid a repetition. The other is the basis on which the modern German economy has been built.

Bismarck's impact on modern Germany can hardly be exaggerated. Bringing together a number of disparate states, whose rivalries were preventing proper development, Bismarck forged the union focusing in cultural, political and economic terms, that became the basis of modern Germany. It was the work of a genius and addressed many key issues. The move towards political union was engineered by sheer political graft, with patient appeals largely based on self-interest, to the leaders of the individual German states. Economically, the route chosen by Bismarck was the forging of a customs union, encouraging free trade within the German states, and an astute application of Frederich List's economic policies, to relationships with the outside world. The second half of the nineteenth century saw the rapid development of German industry and, with it, a growing sense of nationalism demanding a place at the table where the fruits of colonial expansion were being shared out.

Bismarck's period in Germany's social and economic history introduced major and comprehensive social legislation which has been the basis of the general western European approach to welfare legislation. Bismarck's motives were mixed, in the sense that his main concern was to ensure that socialists and communists in Germany could not exploit politically the adverse conditions of capitalist development. Whatever the motive, during the period 1860-80, the new German state introduced comprehensive legislation relating to pensions, sickness benefits and industrial injuries. The belated growth in German nationalism had a downside; the growth of philosophical schools associated with Nietzsche advocating concepts such as the national will and racial destiny; the jockeying for position vis-à-vis the old colonial powers of France and Britain; the inevitable progress towards large scale military conflict resulting in two world wars.

The catastrophic sequence of events, that ended ultimately in the second and truly global war, acted as a catalyst for the drive for permanent peace in western Europe. All participants in the war shared in this desire, but none more than Germany. With its relationships with western Europe destroyed, and, subsequently, control over its eastern European hegemony broken by the extension of Soviet political control westwards, Germany's interest in open frontiers and a restoration of fundamental intra-European relations has been paramount. All European countries seek peace but Germany has most need to use that peace to rediscover and redefine its true national identity. The urgency is revealed in the constitutional debates within Germany. It was, also, reflected in the traumatic events which lead to the precipitous re-unification of Germany after the collapse of the Soviet Union.

Germany's post-war history has been notable for its efforts to promote peace through its commitment to the emerging European ideal. It has also made an important contribution to the way in which the western market capitalist model operates within western Europe. Encouraged by victorious allies, Germany adopted an industrial relations system that has become the role model for state capitalism. Its salient features are a legal framework, which provides for worker participation in the direction of companies, a system of trade unions, which are industrial rather than trade or occupation based, and a national negotiating framework, designed to create what is known popularly as the social contract. The latter term is used to describe a consensus form of negotiation which sets the balance between productivity gains, wage settlements and the background, state provided, welfare structure. Whatever its wider merits, and they remain considerable, this German system contains many interventionist elements, totally consistent with Bismarck's ideal, which has provided a basis for the most successful economic performance in Europe ranking with the performance of other global competitors.

Those who oppose any effort to establish a common European identity usually, first, cite the idiosyncrasies of the British and, as a coup de grâce, so to speak, talk about French behaviour. Certainly there are many elements in the common perception of the British and French characters which resemble a time bomb waiting to blow away any rational approach to building a wider European community. It is strange that these critics do not draw the obvious conclusion that the French and the British are actually so close rather than so far apart. It is commonplace how families argue. Is it not possible that the apparent tension between French and English reflects the sighting of their own image in their counterpart's national characteristics? Shakespeare made France's Charles VI remark, after agreeing to the betrothal of his daughter, Catherine, to England's Henry V, 'that the contending kingdoms of France and England, whose very shores look pale with envy of each other's happiness, may cease their hatred, and this dear conjunction plant neighbourhood and Christian-like accord in their sweet bosoms'. [5] The early settlement of Brittany by the Celts from Cornwall, the invasion of England by the Normans (Vikings) and, in particular, the coming and goings of the hundred years war, suggest a strong element of common stock and roots with at least northern and central France, which cannot be ignored.

France was key to the process by which Europe emerged out of the dark ages and became a driving force in the establishment of a new European civilization. France also benefited from the switch in focus of trading from the Baltic and the Mediterranean to the Atlantic coast and was a major player in building overseas empires - in Africa, America and in the Far East. These points have to be remembered in the task of understanding French attitudes. The French refer to the great period in their history, the enlightenment of the eighteenth century. In this period the French exerted an influence of global proportions, the use of the French language, the contribution to literature and philosophy, as well as political and military influence. French was the lingua franca of the diplomatic and cultural world. Philosophical development, including the logical Cartesian discipline, provided the fuse for the political explosion of its own revolution and continues to influence French thought. The tradition of liberal thinking also spread far and wide influencing movements for reform and liberation, not the least in the newly emerging independent country, the United States of America.

The immediate, and cataclysmic, impact of these ideas took place in France itself. The revolution included the reign of terror, but it was decisive in bringing France into the modern era. It established sound principles of democracy and gave modern shape to its institutions. The process was both consolidated and extended by Napoleon. Like Bismarck, Napoleon was a huge political figure. His influence can still be seen in France and in most western European countries in the use of elements of the legal and political system created by Napoleon.

Within France Napoleon's influence was and remains huge. He instituted wholesale reforms in law, in governance, in church and state relations and in transport. All of these reforms revolved around a centralized political system and the use of a written legal code to oversee most aspects of social activity, including industrial relations. The modernizing and centralizing revolution anticipated the reforms adopted later by Bismarck in Germany. They have made it easier for the two countries, as and when they have desired it, to introduce complementary types of legislation relating to economic and social spheres in later years. It has not been difficult, in the post-war years, for France to adopt a form of the social contract, and methods of worker participation, similar to those in Germany. It also explains the ease which France and Germany share over common ideas over deepening of the European Union.

Britain felt able to defer facing up to its European commitment by seeking to cling on to the past glories of the empire. The dissolution of the empire has taken place peacefully, so much so that many Britons have not fully accepted the reality of the new world in which they live. The fact of being on the winning side in every major war over the past hundred years has also contributed to a sense of complacency. In France's case the reverse is true. Its last hundred or so years has included defeat by the allies in 1815, by the Prussians in 1870, barely surviving a war of attrition, and experiencing invasion, if not defeat, in the 1914-18 war, defeat again in 1940 and an abrupt loss of empire hard after the 1939-45 war. These events were traumatic. So much so that, prompted largely by President de Gaulle, France accepted the need for radical change. It appears to have made the transition from an imperialistic European nation state to a modern society and economy. It appears to have a clear view of its role in a wider regional community, a transition which Britain has yet fully to make.

These three mini portraits of Britain, Germany and France are, hopefully, sufficient to establish that western Europe has created an acceptable market capitalist model resting on quite different cultural foundations from those found elsewhere in the world. The major countries in Europe have a political and cultural perspective in which the role of the state as a partner in economic activity is comfortably accepted. That perspective also accepts the legitimacy of worker demands to consider the provision of welfare benefits generally, as well as the narrower area of wages, in negotiating the contract under which they are prepared to cooperate in introducing productivity gains.

On the wider canvas, the three portraits of Japan, the United States and Europe, have shown that far from there being a universal, culture free system of economic activity, each example of western market capitalism is based on a particular blending of culture and forms of economic activity. The blending is designed to ensure that cultural characteristics are harnessed to give the maximum chance of economic success. It is also a recognition that cultural values are of a higher order and cannot be allowed to be over-ridden by values generated autonomously by economic activity.

Notes

1. There are many good books written about Japan and its economic success, one of which is Vogel, E.F. (1979), *Japan as No. 1*, Harvard, Cambridge, Massachusetts
2. Morita, A. (1987), *Made in Japan*, Fontana, London, pp. 150-151
3. Shakespeare, William, *Richard II*, 2.1
4. Jennings, Ivor (1972), *The Queen's Government*, Penguin Books, Harmondsworth, p. 11
5. Shakespeare, William, *Henry V*, 5.2

Part III
A SOCIETY OF VALUE

11 The imperialism of the economic model

Whatever else is said about Marx, his approach is a constant reminder that economics has to be set in a wider social context. The context underlines that the resolution of key economic issues, such as employment or the distribution of income, depends as much on balances of power within society as on the operation of economic laws. Marx also reminds us that, in setting up a critique of economic activity, of wealth creation, wider issues are at stake than simply the production and consumption of goods and services; economic activity is but one, albeit crucial, arena in which society is seeking to secure its future. Marx also warns that we must be wary of any deterministic argument based on a presumption of a value-neutral process.

There are models of society other than that proposed by Marx. One much favoured by market capitalists is that of Herbert Spencer, whose analysis of society is Darwinian in the sense that it identifies the same selective process (the survival of the fittest) in human endeavour and achievement, especially in economic activity. Spencer's view was very much a child of its age - an example of the Victorian view of continuous progress applied to the development of human beings as well as to technology and science. Similar views have been adopted by later writers.

In his introduction to the 1969 abridgement of Spencer's *Principles of Sociology*, Stanislaw Adreski [1] strongly argues the case that Spencer was an evolutionist in his view of the way societies have developed. It was Spencer, not Darwin, who coined the phrase 'survival of the fittest' referring to the view that a 'system consisting of structural parts whose functions are not adapted to each other or to the demands of the environment will be destroyed by its competitors'. [2] Whilst Spencer did not specifically embrace social Darwinism, i.e. the view that success or failure in economic terms was an example of the survival of the fittest, his views and their application in sociology was too close in relation to social Darwinism for him to be able to escape the label. Indeed, such was his fame and reputation amongst those wanting to justify economic success, that he was given a hero's welcome when eventually he visited the United States in 1882. 'Everywhere he was greeted with reverence by men who saw in their own selection for affluence the strongest proof that the race was being improved.' [3] 368,755 copies of his book were sold in the United States in the forty years following 1860. [4]

Other systemic analyses are to be found in the works of continental sociologists, such as Durkheim and Weber. Weber's thinking, despite its obtuseness, is particularly interesting in arguing that society generates its own particular value sets and that 'subordinate' activities, such as economic activity, are subject to the predominancy of these wider values.

Weber's view of society is as interesting as it is opaque. His is no systematized structuring of society and yet his writings contain a clear view of the interaction between values and social action, or the relation between culture and social life, as Schroeder has put it. [5] This relationship has a number of elements in it. The first is 'the dynamic of charisma and routinization'. [6] This refers to the importance Weber attached to new systems of ideas (often originating under the influence of one particular personality) which have 'a potentially revolutionary impact on social life'. [7] But Weber also notes that eventually 'formerly powerful belief-systems became well-established ... and became integrated within everyday life'. [8] This latter process is what Weber meant by the routinization of charisma. The process is two-way. The new belief system gradually permeates society as a whole, but the system itself also accommodates interests in society.

Weber also makes a distinction between various spheres of life, for example, the political, economic, religious. These spheres overlap and impact against each other, the combination determining the scope for cultural change. In particular 'whenever the spheres of life become more differentiated there is a greater potential for the impact of ideas'. [9] Associated with this view was Weber's belief that the world view, or value set, was able to develop separately from other social forces but there will always be tension between the logic of that development and the lessons it learns as it impacts on the empirical reality of the world itself.

It is clear that Weber was well aware of the potency of ideas, whether they emanated from the cultural or religious element in society or from, for example, economies. He also underlined both the revolutionary, imperialistic, impact of new value systems on other parts of society, as well as the subsequent processes by which the rest of society adapted to the new process and in so doing also modifying them. Eventually the novel becomes established as the orthodox.

One of the objectives in constructing a wider social model is thus to help identify the underlying motivation for behaviour, both individual and social. What is it that drives people personally and socially when they make decisions affecting their daily lives? The answer is as multifaceted as society itself. It depends on personal motivation which can range from greed to the highest form of altruism. It can be the pursuit of happiness or, for some, but only rarely, the adoption of a life of self-denial and service. It can take the form of the pursuit of material possessions or it can be more spiritual. It can lead to a compulsive, aggressive role in society. It can be more reflective, more autonomous states of personality.

What is certain is that wherever people fall in this categorization of aims and objectives, their desire to succeed, and to achieve, surrounds their day to day actions. Mary Douglas' analysis [10] of why people consume suggests that the economic view is limited in many vital respects. It leaves out many explanatory variables. It loses interest in the issue (at the point of purchase) just when the subject becomes interesting to the sociologist. She also argues more generally that sense can only be made of consumption patterns (and, by extension, all activities) by recognizing that they are a veil over culture. Behind these actions are intents, designed to tell people who they think they are and who they wish to relate to. The use of goods and services in this way is, thus, closely related to the task of life fulfilment. It also means that attention has to be directed more widely than solely to consumption. The analysis applies equally to the processes of production, since, as producers, people are also intent on setting up markers as to who they think they are and to whom they wish to relate.

Mary Douglas' arguments, and others like them, add to the debate about the social structure in which economic activity is set in several ways. First, they help to clarify what people are really striving for. Second, they offer a clue as to where to find an answer to the vexed question of why people continue to pursue higher (or at least differing) consumption patterns when the additional consumption is not necessary to survival. The analysis points to a way to understanding wants as distinct from needs. The daily task of life fulfilment, particularly within the context of economic activity is to be seen as a constant effort to achieve identity. That effort may contain many elements of individual satisfaction, but it will be dominated by the social dimension - the need for regard for, and esteem from, others.

The workings of society is as much about means as about ends; this wider view of activity is not stressed by economics. On the whole the economic model assumes that the satisfaction (welfare, utility) to be derived from the activity is as measured at the point of purchase of final goods and services, not before, nor, as we have just seen, after. Nor are economists much interested in the contribution to welfare that is derived from engagement in the process of production, except in the narrow sense that good working conditions or incentives may contribute towards improved performance.

The economic model, certainly the western market capitalist model, is largely built around the need to satisfy personal motivation. In the case of the processes of production, the individual worker is seen as the key contractual party in the labour market. In the act of consumption, market theory assumes that the act is individual and determined by gauging personal satisfaction (utility) largely in isolation from others. Sociologists generally take the opposite view that sees most things as being determined socially. Personal acts only have meaning in relation to others; this difference in approach is fundamental.

The difference between individual and social, between personal and community, has also to be brought in to the debate about appropriate management models. If we take the wider social view, then the motivation to work is best regarded in terms other than the narrow individualistic basis of self-satisfaction and monetary gain. The social view sees work itself as part of life fulfilment (hence the awful tragedy of unemployment) and regards relationships with others at work as crucial in determining the level of motivation and contentment derived from this experience of human beings working in community. It is why it is crucial to get the basis for motivation right in each society. If a society is successful in this respect, as in Japan, a tremendous power is unleashed. If it is not, as in Britain, the ill ease with which the economic and social mix together seriously detracts from performance. The appropriateness of the balance between personal and social values is therefore vital.

How does society establish these values? The short answer is obscurely and with difficulty. The variables are many. It is possible that different value sets co-exist with each other. It is possible that some are applied in one area of society and some in others - and they may be contradictory in essence. It is more likely that the commonest task is to blend together two alternative value sets, particularly to balance the personal and social. The nature of that balance seems to change from time to time and from one area of social activity to another.

What is less clear is the extent to which any permanence can be attached to the holding of a particular set of values. Those concerned with values, theologians, philosophers and perhaps, less genuinely, politicians, like to assume that there are permanent values. However, these absolutes are slippery customers when they are handled at any particular time, and in any particular secular context. Whatever the difficulty, it is reasonable to conclude that it is necessary to look to the wider society for the values that claim priority and which should set the value parameters for other activities within society.

Society's choices, or preferences, of ways of moving forward in certain directions, present an ongoing challenge to all. Does society want to re-allocate resources more to public and away from private uses? Has its preference shifted towards improving, say, the transportation structure rather than wanting more increases in consumer spending? Does it want to spend more or less on the arts? Are its concerns over internal and external national security met adequately by present levels of the armed forces and the forces of law and order? Does it want to increase the amount of investment in education for life as well as in for the under twenty year olds? These, and many more, are examples of questions that society needs to answer to enable decisions to be made, all of which contain an ethical dimension. The ethical dimension, even if revealed, will not suggest a clear cut conclusion - but it will serve to open up a debate about competing ethical choices which inevitably leads to alternative choices as regards ends.

The same can be said about society's choice of the means to be adopted to pursue its objectives. Quite massive changes in the preferences shown for different forms of organization (particularly when they affect the economic/political sphere) take place over time. A current debate is over the role to be played by government acting on behalf of the community. There are many others issues affecting the choice of means. Does society wish the process to be managed nationally or locally? What use is desired of semi-autonomous institutions, specifically created to act in the public good? What role is there for consumer interest groups? More controversially, what conditions does society want to create for, or even impose on, business activity to ensure that certain types of business or management models are followed? The question of workers' representatives on directing boards is one such example; an insistence of the provision of workplace benefits and/or minimum wages is another.

The consideration of these issues of social goals and means is an essential and pervading element in daily community life. As it is so important it is not surprising that power groups strive to dominate this decision making process. Frequently such tussles lead to an usurpation of power. To argue, for example, that many of these choices are predetermined by the economic system is one such usurpation, for it attempts to rule out the consideration of alternatives. Right wing ideology takes for granted that government intervention is bad for business and society. Therefore any suggestion to extend government powers must be bad. The left wing, of course, often argues the reverse in equally dogmatic ways. Or it may be argued, particularly by self-serving politicians, that whilst a debate is needed, that debate can only take place within properly elected democratic institutions such as parliament, thus ruling out, or seriously diminishing, any substantial contribution by other groups or institutions to the debate.

Given these pressures, which in different ways are designed to stifle debate, by closing down means of debate, the business of ensuring that society can provide adequately for an appropriate ethical debate becomes a task of the highest priority. It requires close attention to creating an appropriate forum for debate and to the process by which the debate takes place.

Pride of place must be given to the role of representative government. In a complex society, resting on democratic principles, the role of those elected to represent constituencies is crucial. The institutions of democracy are thus the first point of reference as a forum for articulating and taking political decisions. These institutions must also be given the responsibility for enacting legislation to give effect to choices and for ensuring that an administrative capability is in place to execute the decisions. However, a host of questions remains to be addressed. How much of the debate is to take place nationally, how much regionally, how much locally? What powers are allocated nationally, regionally and locally? What forms of election are to be used?

In a healthy society the needs are for more than the hothouse of parliamentary politics. If it is to have a really satisfactory debate about goals and processes, a healthy society needs to recognise the important contribution that extra-parliamentary institutions are able to make - the heavy structures of earlier chapters. It is astonishing that some can argue that other institutions, such as the churches, the education or legal establishment, or those representing the arts, should have little say in determining society's priorities, except where their cause is taken up by parliamentarians. Such arguments are the special pleading of those with self-interest in parliament or who are ideologically suspicious of the value basis of other institutions. These institutions are admittedly pressure groups, and it is always difficult to devise means whereby due, but not undue, regard is given to their view (the American acceptance of a process of competing pressure groups could teach Britain much). However, it does no good at all, either to seek to exclude them as irrelevant or undemocratic, or to stifle them or to diminish their impact, forcing them to operate only through parliament.

It has been argued that logically the economic model is a subset of a wider social model. The problem in western societies is that the fulfilment of much of the community's aspirations depends in the provision of goods and services. Also lifestyles, and senses of esteem and worth, are largely derived from participation in economic activity. The economic model, therefore, dominates any debate about wider social goals and activities. Indeed, the mode of thought of the economic model is so pervasive that it is often seen as the dominant model, with the theoretically wider social model being forced into a reactive, subservient role.

It is a forgivable view. The economic model clearly dominates people's lives. Moreover many power groups, some ideologically based and some simply pursuing economic wealth, have a vested interest in the model. Consequently, they attempt to control it for their own purposes and seek compliance from others to its alleged rigid laws, which threaten personal and social ruin if defied. The claim of its supporters for the system's scientific neutrality is also a powerful argument. If this claim is accepted, much of the legitimate protest about the workings of the system is neutralized. In a system totally governed by scientific laws it is pointless to ask whether the outcome is just or not. Events will be what they will be - what the laws dictate. Abide by the laws - success. Buck the laws - disaster. There is no place for the cry, 'it's not fair'.

This argument is nothing other than counterfeit. The western market model, or any other economic model for that matter, is not value neutral. The recognition of that reality must permit the consideration of alternatives, of offering the double benefit; alternative value sets for achieving economic success, and an economic system more consistent with wider social values.

Many suffer under this form of economic determinism. They include those who lose their jobs as a result of technological change; those who are forced to submit to processes which demean their humanity; and those who would wish to operate on the basis of other value sets. But there are many who feel they are winners in the game. The dominant feature of the winners - and the term is used to describe the majority of people who believe they gain from participating in the modern form of capitalism - seems to be a restlessness and a dynamism revolving around an insatiable appetite - to consume and to work.

A strange feature of the history of western capitalism, which has seen absolutely massive productivity gains, has been the fact that such a small proportion of the benefits of productivity growth has been taken in a shorter time at work - whether in the form of shorter working days, shorter working weeks, longer holidays or earlier retirement. Most of the gains (of increased productivity) have been taken in a higher material standard of living - more goods and services and, therefore, little let up in the pressures of work.

A characteristic of the market economy is the loyalty of management (if not the workforce in general) to the institution they work for. Those working in a company have a direct interest in its success. Successful endeavour guarantees employment, provides promotion and higher earnings. The objectives of a company also include securing its future. Sometimes this means accepting decisions that can be detrimental to the interests of the present workforce, such as major redundancies to shake out an underlying lack of competitiveness. It requires a strong will to accept decisions of this sort, detrimental to themselves, but for the benefit of future generations, of workers and shareholders. Company loyalty built up in order to compete successfully, and the strong desire for the fruits of success, are good examples of the strange dynamic of western capitalism. It results in an atmosphere built largely around a fear of competition - especially competition perceived to be coming in the future - that never allows the conduct of the business of the company to relent. The next stage of productivity gain must be brought in as quickly as possible, marketing plans must be prepared with extreme urgency; potential sales must be pursued to exhaustion. The result is an atmosphere where long hours, physical inconvenience, masochistic team-working prevail. The permanence of perceived competitive threats dominates and drives management so that absurd extreme and adverse working conditions are also accepted as permanent. Most workers appear to accept these very high costs and adverse working conditions. It says much about the strength of the desire for more material gain and the desire to achieve.

Although in many respects company management is dominated by perceived threats of present or future competition, this competitive element of economic activity runs in parallel with the equally strong cooperative view of appropriate ways for people to work together. Most management text books stress the virtues of team-work, of appropriate leadership and of appropriate motivation. The military model gives a clear instance where the competitive and the cooperative must co-exist. The enemy in a military engagement is a real competitor - engaged literally in a life and death struggle. The response to this threat is equally clearly regarded as cooperative. A fighting unit needs to support its members, to help each other, to sacrifice for others if needs be. The two elements are clear. There is an enemy to be defeated. An interdependent highly motivated team is needed to do the task.

The grounds for transforming the same imagery over into business activity are not wholly convincing. In the first instance, the nature of the enemy needs to be examined carefully. The market capitalist system claims to meet wants at minimum resource cost. The way the resource cost is kept to a minimum is by allowing competing suppliers to 'do battle' in the market place; that way the efficient will win over the inefficient. There is some truth in that proposition that must never be lost, but the principle is used to justify a constant state of fear about the way other corporations or nations are planning 'to defeat us'. There appears to be no end to the list of potential competitors (Japan today, South Korea and Taiwan tomorrow, China and India and Russia, yet to come). People are allowing themselves to be trapped in a hysterical spiral which defeats the true purpose towards which the provision of goods and services should be devoted, namely, a satisfying and responsible lifestyle. Moreover, it is clear that this obsession with competition (no longer in the strict economic sense but now as a battlefield) inevitably rubs off on the models adapted within companies to ensure motivation and performance. The respective efforts of managers are allowed to become competitive; let two rivals for promotion fight it out; the nasty boss is reluctantly admired for getting things done; fear is accepted as an element in motivation. The ultimate criticism of this obsession with constant competitive threats is that it has no end in sight. There is no final success. A year of intense pressure at work is rewarded not by relief but by a year of even more intense pressure. The company's success in seeing off a competitor simply leads to another battle with another such competitor.

This compulsion to produce efficiently is reflected in longer hours, stress, hurtful labour force rationalizations and unremitting pressures to perform. It is also parallelled by a similar compulsion to consume, a similar hyperactivity attached to people's activities as consumers. People are driven by advertising, by the stimuli of the walk down the high street. Their shopping list of goodies, for the home and for pleasure, constantly outstrips their ability to purchase. The desire for goods dominates the lives of many who desperately want to take the 'waiting out of wanting', as the early credit card advertisement put it. Reference has already been made to Mary Douglas' analysis of why people consume goods and services. [10] The sociological analysis helps to answer the question 'will, can, the compulsion to consume ever be made to slow down?'. The slogan 'enough is enough' was coined to refer to the blasphemy of a rich north co-existing with a poor south, but the question also needs to be internalized within northern hemisphere capitalist countries.

Another issue, raised by the market capitalist model, is the way in which its supporters now seek to apply its approach and its value basis to other areas of social activity. The market capitalist model is unsurpassed for its sheer aggressive vitality. Not content, it seems, with what it regards as a total victory over other economic systems, it appears unable to resist the temptation to invade other areas of society in order to bring them under its iron rule. The area of business activity, where once it was argued that ethical and moral issues do not apply, is now praised for the strong ethical/moral basis of its activities. It is not simply that business activity is now accepted as an arena for practising personal virtues. It is also seen to be proof of the success of living by libertarian views, of self-reliance and independence. Moreover, the market system itself is alleged to be an essential concomitant to democracy. The market place needs a democratic system in which to operate but the market also helps to preserve democracy.

The unapologetic ethical justification of market activity, in both the narrower personal and the wider social sense, is a reflection of the degree of confidence now enjoyed by supporters of a libertarian view. So too is the way in which the virtues of the market place are being advocated as the appropriate framework for managing other parts of social activity. It is a troublesome development in that, like the emergence of the monetarist debate itself, there are a number of different levels of argument. At one level the issue is simply put forward as one of resource allocation. All the major areas of state spending involve the use of resources. Since no-one wishes to waste resources, why not use the same tried and tested market model, which is designed to address this precise problem? Hence the introduction of the familiar policy changes designed to create market conditions wherever possible; identify the product/service, identify the suppliers of the product/service, identify the purchasers of the product/service, create market conditions to produce the most efficient use of resources. Even at this level many questions are raised. Is the market transaction an appropriate way of addressing, say, the provision of care? Or of legal and education services? Is there not a risk of destroying the essential nature of the transaction by regarding it as marketable? The debate becomes even more complicated (and angry) if there is the suspicion that the advocacy of market principles is being used merely to justify cutting the absolute level of provision of services.

The problem goes deeper. The libertarian view enshrined in the market has some strength. Prizes should go to the winners, the industrious, the deserving. Exposure to market forces is critical in forging character and developing self-reliance. However, this view, when carried over into other spheres, runs the risk of prejudging many of the issues. Its worst aspect is that the justification of business activity contains a distinction between deserving and undeserving, since it assumes that success comes from self-effort and failure from an inadequate level of self-effort.

It has to be asked, however, whether carrying this distinction over into other areas is an appropriate way of addressing many of the non-economic issues facing society? It clearly has limitations in the education sector. It gets itself into difficulties in trying to make a distinction, or failing to make a distinction, between poverty and illness and criminality within the legal system. It also prejudices the approach towards income maintenance. As was argued earlier, it also provides a dangerous form of moral release to those who hold the libertarian view. If success is there for the self-reliant and industrious to grasp then what is to be said of the 'failures' in society? They may be failures because failure derives from force majeure, or fate. Many Africans suffer from being born in areas where drought and desert are eroding cultivable land. The disadvantaged, physical and mental, suffer from cruel fate. Libertarians can sympathize with such disadvantaged and, according to their means and the sensitivity of their conscience, give charitably to help. But they need not regard themselves in any sense as to blame for the condition of such people. If blame is to be attached, it should be directed at God or fate or whoever.

The unfortunate may also be failures because, in the libertarian view of things, they have not tried hard enough - to stay in a job, to look for a job, to spend money wisely, to live at home with their parents (in the case of young people), to overcome their alleged tendency to laziness in the case of some minorities. In the case of all of these categories, to a libertarian it is quite clear where the blame lies - it lies with the failures. They are to blame, not the industrious libertarian. So the libertarian view provides a very convenient release from responsibility for others.

The contribution of the libertarian view to the underpinning of the market capitalist model and to wider society, cannot be denied. The role of self-endeavour and effort in earning one's living is very important to success and the economic aspects of such endeavour are also very important in terms of life fulfilment. The same approval, and the expectation of similar results, will have a wider application in society generally. But the present way in which the articulation of the libertarian view within the economic model is being encouraged to cross into so many other areas of society is a form of imperialism.

There is a different view. Different value sets may be more appropriate in other areas and boundaries may need to be erected to defend them from libertarian attacks. It may be desirable to ensure that opposing value sets can co-exist. This view, however, has had very little success in Britain in resisting the contemporary libertarian onslaught. Part of the reason for the poor defence has been the demise, as Dennis and Halsey [11] might put it, of the Tawney man. The half century ending in the 1970s, saw the flowering of an alternative to the libertarian view to personal life fulfilment and to social development in a market capitalist society. Its characteristics were an emphasis on a cooperative model of working together, distrust of unbridled capitalism, and a desire to constrain rampant consumerism. The view was encapsulated in the idea of the Tawney man - a common man, a common sense man, a cooperative man, a responsible man [12]. Put such a man at the centre of things and all would be well. Production would be participatory, responsible and therefore more efficient. The fruits of the production system would be used more wisely and responsibly. Adequate care and compassion would be provided to the victims of the complex industrial system, principally by the state. The Tawney ideal disappeared from sight in the 1970s, a cause of great concern for those who believe society needs a dialectical debate.

The only sensible starting position for a critique of the relationship between economics and society is the fact of change. As has been argued, by its nature, the economy is always on the move. This phenomenon of change, and its consequences, has been most vividly apparent in the period beginning with the first industrial revolution and which continues today. Economic change generated by technology is constantly forcing society to struggle to adapt, to react and to try to retain control. The power of technological change, and its resultant economic dynamic, cannot be overestimated. Chapter Two illustrated the process of change from the viewpoint of one such technological development - the introduction of steam power. Many other examples of new technologies, being introduced at the same time, could have been examined; new technologies in textile manufacturing, iron and steel manufacturing processes, new transportation infrastructures, particularly canals and roads, and many others. The process is a dynamic one and the introduction of new technologies has continued relentlessly, the only distinguishing features being the existence of alternating periods, when the rates of introduction of new technologies have speeded up or slowed down.

Changes in technology are partly determined by economic circumstances. The result of research to solve a technical problem; to meet a hitherto unmet demand or to respond to a crisis such as the depletion of existing raw material supplies. Much technological innovation comes increasingly from activities such as defence and space exploration where the technical challenge is not directly economic in origin, but, when solved, turns out to have economic consequences. The production of microcomputers, for example, dictated by the weight and space constraints in space travel, contributed to the introduction of the current phase of computer technology. Technology is harnessed, sometimes, to achieve a clearly defined outcome in economic terms and, sometimes, to create innovations for which the ultimate uses are less specific. In either case, the process put into motion by research and development is invariably open-ended with little realization of the economic consequences, still less the social consequences. Chapter Three pointed out that a feature of new information technology is its novelty in the sense that its subsequent exploitation has not been predicted (perhaps was not predictable). Semi-conducter devices are capable of pervading a whole range of economic activities, substituting, developing and radically changing existing ways of producing goods, providing services. Its results are massive economic dislocation, changes in working practices and in lifestyle, the social consequences of which are only just emerging.

The fact that the economy is the prime innovator of change gives it a degree of control over society which would not be justified otherwise. It is also reinforced by (although it can be argued whether this is a separate issue or simply a consequence of the first point) the dominance of the consumption of material goods and services in the pursuit of most people's immediate goals.

Since the economy is the principal source of change it is thus able to exercise a form of imperialism vis-à-vis the rest of society. Those driving the economy, as it were, bombard the rest of society with new things, with new processes, which give the economy an initiative that can hardly be resisted. This imperialism seeks to use its power, and control over events, to further its interests. The economy has its own culture, its own way of doing things, its own model of understanding of the way things work and its own value system. Its primary need is therefore to defend itself in these areas. Whether this attitude is summed up in the tag, 'business is business' or is seen as a much subtler and deeper defensive strategy, involving the control of political pressures, is a matter for debate. However, the intensity with which the economy is prepared to defend itself, is not.

This imperialism is also reflected in the impact that the output of economic activity (goods and services) has on the rest of society. The continuous flow of new goods and services, with the challenges they offer to conventional ways of doing things, of deriving satisfaction, inevitably pushes society, at large, off balance. Society is in the hands of the economy, and society can only react. New forms of transport change leisure patterns, new forms of industrial organization change travelling to work patterns, information technology in the home changes leisure activities, longer and more frequent holidays affect traditional patterns of life such as regular church going (surely a socio-economic change rather than anything to do with faith and belief?). Imperialism dominates and that is exactly what the modern economy does. Its weapons appear too powerful to be resisted.

The imperialism is also the way in which the economy, full of belief in itself, seeks to apply its principles and practices to areas outside the natural definition of its territory. It is to be seen, in the modern context, in the debate as to how best to manage other areas of society, the provision of health, education and social services. It is not simply that, since these areas use resources, they should be regarded as a part of the economy. It is the deeper issue of presuming that the relationships, 'transactions' undertaken between suppliers and consumers of, say, care services, should be regarded as the same as transactions between suppliers and consumers in the conventional economic sense and therefore subject to the same disciplines. The distinction between the concept of a 'transaction' compared with, for example, a 'gift' goes right at the heart of the current debate over the management of social activities that use resources.

The imperialism also seeks to impose its ideology on others. A prior question is whether an ideology based on individualism and independence is the only value basis for market capitalism. The attempt to export an ideology, such as that underpinning market capitalism, to other parts of social activity may be an extended defensive mechanism to preserve the economic model's freedom to act in its own way. But it may be part of a wider ideological debate to promote the economic model's ideology over alternative value sets in society. All these are complicated but absorbingly pertinent questions. Whatever the motive, the economy and its ideological base, is a highly active ingredient in this debate; so much so that it persistently puts the rest of society on the defensive.

Congenial or not, most social institutions are thus forced into a reactive role in the face of this militant onslaught. It is partly because many daily issues affecting the conduct of social institutions appear to be primarily economic questions. In debate about changes in the schooling structure or the way the police protect the community, the case invariably revolves around the cost of alternative proposals. Debates about the availability of resources or arguments based on value for money are legitimate but they create a climate which prevents a debate about non-economic issues. The dominance of economic criteria is also partly related to the fact that technology, and its application, is constantly offering new ways of doing the work of social institutions. The use of television and computers in school is a prime example. Another is the use of electronic tagging of people on bail or of genetic fingerprinting in the detection of crime. No-one can deny the value of these, and other innovations, to the work of social institutions but they again serve to put social institutions on the defensive.

The value-for-money argument is an even more decisive intrusion into the way social institutions run their affairs. Social institutions in the main are effective according to the amount of resources they have at their command. These resources are not unlimited and debate is needed to determine the level of resources to be provided. Economic considerations dominate. They have the larger task of creating and allocating resources as a whole and it is natural that an attempt should be made to replicate the same conditions found satisfactory in the economy proper (i.e. the operation of the market) in other areas wanting resources. Hence the ease by which the economy has retained the responsibility for determining the best use of resources, and for the use of the concept of the transaction in determining the allocation of resources. Social institutions have been unable, and maybe should not be able, to find a way of entirely preventing the debate about their task being conducted in this way. But, unable to deny the importance of giving a proper consideration to resource allocation, social institutions again find themselves frustratingly in a reactive situation. The use of the transaction concept, justified on somewhat narrow terms in an economic context, has, however, other implications since it opens up a debate about which ideologies other social institutions prefer to adopt as a basis for their actions.

Change is, and should be, a reality in the conduct of social institutions and change may, from time to time, require a review of fundamental values. However, the health of society requires a much more robust articulation of the values and role of social institutions than that provided by the reactive debate just described. If such a wider debate is conducted it will allow society to distinguish properly between ends and means. The ends pursued by social institutions must be based on the values which society at large wishes to apply to these areas. The means by which these ends are pursued may, or may not, be based on the same values. Again, that will need to be determined by society at large rather than allow them by default or, by force majeure, to be determined by the economy. Social institutions must delve deeply into the sources of the values so that they are ready and able to take an active part in the debate about society's ends and means.

The need for this wider debate has never been greater. Its conduct, for a variety of reasons, has been hampered throughout the west, in recent years, by the preoccupation with economic regeneration and the particular form that has taken. In Britain's case, it has also been hampered by the narrow view taken on the legitimacy of various players in society in the process of that debate. The origins of this narrow view are both long and short. In the longer perspective, Britain has built up a form of parliamentary democracy whose stability is the envy of the world. Included in this tradition is the concept of parliamentary sovereignty. The supremacy of parliamentary sovereignty can imply that other expressions of, what one might call, public opinion, are either illegitimate or flawed as reflections of the popular will. But this view has resulted in a system in which the representations of special interest groups have always been regarded as suspicious because they are not democratic, an underlying view that has been reinforced dramatically, but, hopefully, only temporarily, by the reaction that followed the disastrous experiments in community action in the 1960s and 1970s in Britain.

The seizure of control in many inner city local authorities, and the subsequent abuse of power by interest groups or political cabals, provoked a political reaction in the form of a very sceptical view of the role and legitimacy of representative groups other than Parliament. As a result, local government has been stripped of many powers, and national government has been unwilling to listen to, or to enter into debate with, non-parliamentary bodies at large - ranging from the Confederation of British Industry (the employers' organization), to the Trades Union Congress (the labour organization), as well as bodies representing consumer interests and bodies concerned with the value base of society such as the churches.

The result of this political reaction has been to damage community relations badly, creating a need for a new institutional framework in which proper and adequately wide debate can be resumed. The case for such debate does not rest on eliminating or constraining the arguments based on the value sets which appear to be popularly accepted at present, but it does mean allowing others a legitimate say. In particular, the conditions need to be created in which the bone fides of all relative groups, including those not part of the mainstream democratic processes, is accepted.

Many forms of ethical debate of this nature can be tried. Exploring the concept of civil society, in particular, deserves particular attention. Amongst many advantages the concept emphasizes that, despite their important role in the process, parliamentarians have no exclusive say in determining society's goals and the means used to achieve them. A new structure needs to have a local, immediate, context, despite the unfortunate experience of the 1960s and 1970s. The pursuit of an alternative structure has not been helped by the reduction in local government powers over the past fifteen years brought about by central government action. Wider social structures will also be needed and the better post-war experiments, including the national planning framework introduced by the Conservative government in the 1960s, might well be re-examined without apology. Above all, the paramount need for an ethical dimension to be given to this debate needs to be recognized. Not necessarily to unify the ethical base but to acknowledge that the fundamental choices before society are always ethical and need to be addressed.

There is an essential value judgement at the heart of the intricacies of social behaviour, where most activities are the result of a complex mixture of competing pressures. It is to be found in the way in which society formulates its aims and objectives and also in the way it allocates its priorities. It is also to be found in the choice of means that are used to achieve those objectives; this is true of society at large. It therefore must govern behaviour in the more limited area of business activity. The way business is done, the choice of management models, the relative importance of different ways of enjoying the fruits of work and the many choices made regarding the appropriate form of institutions, can all be traced back ultimately to a value judgement. It will indicate the degree of preference to be placed on individuality. It will also indicate the importance attributed to community values. The degree of success that can be attributed to society's activities will largely reflect the ability of society to apply its value sets consistently. Society will achieve its goals more effectively where there is consistency than where there is not.

Values underpinning society differ from one society to another. The values underpinning Japanese society at large are quite different to others. The difference is also reflected in the value sets on which economic activity is based, and it has led to a communitarian approach. The benefits that Japan draws from the consistency between the value sets underpinning economic activity and those underpinning society at large can be seen clearly. In other cases, such as Britain, there is a fundamental inconsistency at present. The British, like the Japanese, place emphasis on communitarian values; they have a strong interest in community; they like to work as a team. But the ideology underpinning British management approaches, because of their American origin, derive from American culture with its greater emphasis on individuality. In these circumstances the inconsistencies grate against each other and prevent the full unleashing of effort that flows when there is consistency.

There is nothing static about these value sets, they change continually over time. The appropriate models for economic activity were quite different. For example, in the late seventeenth century in Britain. Each period faces the same task of achieving consistency between the economic and social values. Post-war Germany required a different approach to Bismarckian Germany.

The choice is therefore an ongoing one. It demands a constant reappraisal of how society wishes to go about the business of living and of earning its living. In that constant examination, some values may appear to have deep and unchanging roots. In other cases the need is for a swift adaptation in order to utilize less obvious personal or national characteristics. The need for choice underlines the need for an act of will; there is nothing predetermined in the matter. Far from it, society's freedom and ability to make such choices, emphatically forms part of the business of life fulfilment, where people decide individually and socially what ends and means will most satisfy them at any given time.

Society contains many elements; there are many different objectives to be pursued and many different means of pursuing them. There is more than one set of values underpinning these activities. In some cases, certainly in social welfare areas, most societies seek to apply communitarian values. They are the practical and institutional application of neighbourliness, activities which appeal to a degree of altruism. Other areas, such as the pursuit of the arts, may embody individualism, designed to give as much free expression as possible. They may be passive and reflective. In other areas, such as business activity, values may be aggressively individualistic or rely on a goals-related communitarian approach. These different value sets may be complementary or they may compete. The experience of recent decades suggests that society has deluded itself into thinking that most are complementary - that a contract exists, so to speak, under which certain value sets are used in one area and others in another, with an agreement not to compete; that is certainly not wholly true; it may not be true even generally, for it disguises the fact that there is a constant battle between value sets, especially where the use of resources is concerned, between those involved in producing the resources and those wishing to use them. The frontiers between these two change constantly. The view that values are complementary is also dangerous in that it assumes that one and only one value set is appropriate as the basis of action in a particular area of society. To make this assumption can reduce dangerously the degrees of freedom available to society by which to encourage human fulfilment. It is also a grave misjudgement, and can only serve to diminish the potential for improving the quality of life, to assume that the value choice between personal and community is the same as that between economic activity and other social activities.

Notes

1. Spencer, Herbert (1969), Adreski, Stanislaw (ed.), *Principles of Sociology*, Archen, Hamden, Connecticut
2. Adreski, Stanislaw op. cit. pp. xxviii-xxix
3. Galbraith, J.K. (1977), *The Age of Uncertainty*, BBC/Andre Deutcsh, London, p. 46
4. Galbraith, J.K., ibid. p. 45
5. Schroeder (1992), *Max Weber and the Sociology of Culture*, Sage, London
6. Schroeder, ibid. p. 9
7. Schroeder, ibid. p. 9
8. Schroeder, ibid. p. 10
9. Schroeder, ibid. p. 10
10. Douglas, Mary and Isherwood, Baron (1979), *The World of Goods*, Allen Lane, London

11. Dennis, N. and Halsey A.W. (1988), *English Ethical Socialism*, Clarendon Press, Oxford
12. Those concerned with correct language will, I hope, forgive me this lapse into masculine terms in the interest of rhetoric!

12 Britain and the challenge of European Union

The current moves towards greater union in Europe in general, and Britain's attempts to identify its own European role in particular, represent a pertinent example of the challenge posed by this book. They bring to the fore the need to understand what the underlying pressures are pushing towards greater union. They challenge the sense of history in individual countries as to where their future lies. They open up the debate about whether the objectives of the Union are purely economic or have a wider political and cultural context. They expose issues about the political form the Union should take and, in particular, the degree to which countries are willing to concede sovereignty. They also raise the question of where to draw the frontiers of the new Europe - a question that has been given particular pertinence with the demise of the Soviet empire. Above all, they raise the question of what is meant by a European identity. What sort of region is being created and how will it relate both to other equally advanced trading partners and to the rest of the, developing, world.

Someone has aptly pointed out that Europe consists of 'three hundred and twenty one peoples sharing only American culture in common'. Events in the former Yugoslavia and the threats of similar disintegration in some of the components of the late Soviet Union underline the still inherently fragmented nature of Europe. The paradoxes of cultural history in Britain, Germany and France, have already been revealed, where the conclusions to be drawn from the past are at best ambiguous. And yet, the period since the end of the second world war, has been marked by a persistent and largely successful drive towards creating a greater sense of unity and turning those ideals into practical terms.

The primary motive behind the creation of the European Community was to ensure peace and avoid a repetition of the devastating global conflicts of 1914-18 and 1939-45. The key to that peace process was seen to be, first, the need to lock France and Germany together in a political framework and, secondly, to neutralize the competitive economic interests perceived to have contributed to the causes of war. Whilst the primary practical interest of the Treaty of Rome of 1959 was to create a common market, wider goals were set out in the Treaty aiming for an eventual total political union.

The emphasis on economic questions also allowed the continental European partners to address an additional problem. Post-war Europe was totally devastated by the 1939-45 war. The prospects for countries finding a way individually to recreate a viable economy and to resume as players on a world scale (and, in particular, compete against the United States) were low. Creating a common market was thus intended to create an internal market large enough to provide the economies of scale required to provide the appropriate level of global competitiveness. Although from the beginning, the European Community participated in worldwide moves towards greater free trade, through the activities of the General Agreement on Tariffs and Trade, the prior requirement was the recreation of economic power within the region; if necessary, behind trade barriers with the rest of the world. The prime example of this dual policy, of creating internal free trade with external tariff barriers, was agriculture; an industry, too, where economic policy was formulated in the context of a wider social and cultural background.

The objective of a smooth working internal market has been generally accepted by all mainstream political traditions amongst the member countries. That acceptance has been shared by the British right wing parties, whose principal interests have been limited to creating greater free trade; but they have been surprised and embarrassed by the form progress has taken. If a free market is to be effective it must ensure the free movement of goods and services and factors of production throughout the region (a level playing field, to use the popular terminology). In the case of goods, the objective is relatively clear and can be achieved by reducing tariffs which otherwise distort prices. Even with goods, the implication of creating a level playing field quickly becomes confused. There are many non-tariff barriers that also have to be tackled such as a pseudo concern for quality, for safety, or for environmental reasons, which in reality are discriminatory. Italy's objection to the use of soft wheat grain to make pasta, or Germany's insistence on a high meat content in sausages, or Spain's pressure to increase the required orange content of marmalade, are argued in the interests of common grading so that like is compared with like. In effect they are attempts to protect national interests.

There is a logic in the level playing field argument that requires these issues to be addressed. The problem is magnified enormously when similar considerations are applied to services and the movement of factors of production. The level playing field argument requires that labour can move freely within the community. An easy concept to apply for semi-skilled or unskilled workers but to be applied comprehensively throughout the service sector requires ultimately an acceptance of common professional standards for doctors, accountants, lawyers and others, so that these professions can practise in whatever community country they wish. Transport is another area where the level playing field issue leads ultimately to determining what size lorries can or cannot drive on European roads, and restrictions on the number of driving hours permitted. If unit labour costs are also to be comparable, common approaches to social benefits are needed.

The mass of activity created by applying the level playing field criterion, and the growing perception that much of it intrudes on national sovereignty, is hard to accept by those who, whilst insisting on pushing free trade to its limits, wish only for narrow economic benefits.

The original agenda of the European Community did not have to embrace matters other than the economic in order to raise quite fundamental issues of principle of what the Community is all about. However, the extension of the argument, beyond the purely economic, was already envisaged in the Treaty of Rome. In the consequent developments, of Community policy, Community institutions and increased political integration, members have been encouraged, or forced, to respond to the exigencies of closer union in its wider sense.

The massive step towards economic integration, taken in the Single European Act of 1992, effectively put in place the measures necessary to create the level playing fields so desired by everybody. Another major move forward, on both the economic and political fronts, was the Treaty on European Union, popularly known as the Maastricht Treaty, signed by heads of government on 7 February, 1992. The economic provisions of Maastricht maintained the momentum moving from a common market to common economic policies; and, ultimately, to common financial institutions including a common currency. The political side pursued the goals, already envisaged in the Treaty of Rome, to complete the process of economic integration with political union. The political initiative now raises, understandably, the issue of the degree to which, the speed towards which, community members move in adopting common foreign and defence policies.

Adverse reactions to these unifying trends set in train by the Treaty of Rome, can be found in all countries and not only Britain. They reflect the schizophrenic view that citizens of most European countries have about national identity. But the die is now cast and the contemporary debate is not about will there be a European Union but about what type of Union is to be created. The issue has been complicated, but not in a damaging sense, by the need to include the issue of widening, as well as deepening, the community so that the present partners can find a way of relating to the eastern part of the continent.

The movement towards greater European Union has gathered momentum over the past years and issues of crucial importance are occupying the agenda. The paramount need is to ensure that closer union will guarantee that economic gains will continue to flow. Improving the rate of economic growth was an original community objective and the success in the 1960s provided the stimulus for the community's development and enlargement in the 1970s and 1980s. The need for the region to continue to be successful in economic terms remains crucial. If the pace of growth falters, as it has over the past few recessionary years, and people's standards of living fall (and jobs lost), the environment, against which future developments will have to be considered, will become soured.

The process of creating a common market is not complete. Rationalization in the motor industry and the information technology and communications industries, is needed in order to create corporations capable of being effective worldwide players. One-off gains, arising ultimately from the Single European Act of 1992, have been estimated to be a two per cent reduction in total costs due to removing barriers leading to a one-off gain in output of between five to seven per cent, a fall in price levels of six per cent, together with an increase in employment of two per cent. Gains of these magnitudes need to be repeated as the unifying process continues.

The factors offering future gains will continue to be those reducing obstacles to trade (tariff and non-tariff). Gains will also come from developments whose future form is less certain but are likely to be positive. These include the integration of economic and monetary policy, and opening up the market for services, especially in the financial sector (where the protection these industries have enjoyed for a long time, suggests sizeable economic gains to come from more competition).

The gains will be both internal and external. Improvements in productivity will contribute to continued growth in real incomes within the Union. Like the United States of America, the size of the internal market will be such as to provide sufficient competitive stimulus to European firms. The region will also become stronger vis-à-vis the rest of the world with less reason to fear competition from outside.

The moves towards economic integration (the original treaty, the Single European Act and Maastricht) have created a number of tensions, which are reflected within and among Union members. Some continue to see the objective of the European Union in economic terms; to reduce trade barriers to improve economic performance to make the Union better able to survive as a player in a global free trading area. The creation of free trade within the region is seen as a stepping stone towards participating in global free trade. The view implies a commitment to the principles of free trade both for Europe and for the world as a whole. It also implies that the Union must be prepared to compete in the world market place against all comers. This view is in the tradition of Adam Smith; the economic gains to be achieved are those that exploit comparative advantages in the context of world trade.

Others hold a different view; that opening up the internal market is necessary to improve productivity. Further economic gains continue to be a goal for the Union. An open internal market is also acceptable because most European nations engage in economic activity in roughly the same way or, if there are competitive differences, individual Union members would be able to adapt without any dire or unacceptable consequences to social values or culture. Particular cultural conditions that may be established, such as some elements contained in the Social Chapter of the Maastricht Treaty, indicate the constraints Europeans want to impose on the ways of conducting business or the use they wish to make of past productivity gains. These constraints are expected to contribute to motivation and increase productivity. They also declare an identity of interest in social decisions as to how to take the fruits of extra productivity; in, say, increased safety at work or longer holidays rather than in even higher consumption of goods.

Trading with the rest of the world according to this view would be governed by the requirements of balance of payments policy, by a need to exploit technical innovation arising outside the Union and by the need to cooperate in efforts to increase the growth in world incomes especially in the developing world. These objectives may be pursued by lowering world trade barriers. But a commitment to freer trade has to be conditional. The condition is the need to recognize that there are social and cultural limits to the extent to which the Union is willing to act in the cause of free trade. For example, under global free trade, severe penetration in Union domestic markets could occur from imports from a country where, say, the work culture included conditions of working unacceptable to Union members. If the only riposte to this threat in free trade terms was to adopt a similar work culture, that response could be refused and other ways of reacting including, if necessary, import control, legitimately considered. Similarly, if the European Union has a higher regard for safety on the road than other countries and therefore enacts more safety legislation (concerning lorry maintenance, length of driving hours, etc.), Europe has a right to put the preservation of that social legislation above the narrow considerations of cost advantage gained by a country without the same social view.

The above point of view leads to a form of social and cultural preservation by limiting the degree to which the consequences of free trade are allowed to be felt. It is held in the confidence that opening up the region's markets to its own members, risking no or little such social or cultural clash, creates a large enough market to ensure adequate pressures for economic efficiency, adequate stimuli to research and development and adequate scope for increased productivity.

Such a view has connotations with the Listian infant industry argument, and it is, to some extent, similar. But it is a much wider application of the underlying principles of List's approach. It acknowledges that national self-interest, in this case a regional self-interest, about wider social and cultural values, must be the determining criterion. It is a view that makes sense of much of what is proposed by Britain's partners in the Union when economic and social policies are proposed, in agriculture, industrial welfare or in management models. A similar view is also to be found amongst the more sceptical authorities in the field of economic development, who oppose the pressing of unadulterated free trade policies on developing countries regardless of the cultural conditions of the country concerned. [1] The concern is an application, on a regional scale, of the point made in earlier chapters about the need to identify clearly the ideology upon which economic and social activity is to be based.

What ideology is the appropriate one for Europeans in the twenty first century? The choice will not be self-evident, as the debates in the Union about the ways of going forward have amply illustrated, but the reality and the importance of the choice is in no doubt. The majority of the Union appears to wish to select from its history those elements that favour a communitarian base as the building block for the new Europe. The consequence of this choice is that the region will use its industrial power to defend its particular European ways of doing things in economics and business. It is likely to choose benign cooperative management models with an emphasis on participation. It is likely to be content to see more of the fruits of increased productivity taken in better working conditions and in less intensive working hours. It is likely to seek to retain control of the impact of economic activity on the region's environment. It is likely to want a say on the quality and consequences of consumption, both public and private.

It is also likely to pursue a form of social contract which would be designed to exploit those community values upon which economic activity should be based. In that way economic activity would be conducted in a way faithful to wider social values. Such constraints would be applied in the belief that they would evoke the optimal effort to produce resources efficiently and effectively. The formulation of that social contract would involve government, management, employees and consumers throughout the whole Union. It would also need an ethical dimension.

There is a logic determining the institutional development of the Union which stems from the original Treaty of Rome. The basis of the Union is federalist, not confederalist, and events are driving the Union steadily towards the features of a federalist Europe; a common central bank, common currency, common passports, common legal systems, common foreign and defence policies. That this means giving up sovereignty should not be seen as threatening. All cooperation between countries, all treaties, imply a ceding of sovereignty in one form or another. The reality is that the politics of the modern world have substantially reduced national sovereignty already. Britain is not sovereign over its defence policy. It is part of NATO and is committed to corporate action and not going it alone. Even France, whilst not a formal member of NATO, applies foreign and defence policies largely in conjunction with other powers. To act alone or unilaterally would be inconceivable and would be an act of folly. Nations also delude themselves if they believe they are free to manage their economy unilaterally. Monetary and fiscal policies are now largely dictated by the global economic and monetary climate, such as international interest rates, the fiscal stance of the USA or Germany. It could be argued that the current problem is that, having been forced to yield sovereignty so far, and having lost whatever gains nations felt they had from full sovereignty, the only choice now is to create a regional sovereignty; to replace the loss of national sovereignty and provide a defence against international forces, that otherwise would overrun the nation state completely. In mid-stream there is neither control over the monetary system that nations think they have nor the strength of a regional union to defend its members in times of trouble.

Movement towards a union, based on a sharing of sovereignty at regional level, is to be welcomed. This trend is not the same as determining the appropriate levels of government. In many ways the exercise of national government is now an anachronism. The principle that political decisions should be taken at the lowest possible level is a sound one - the principle of subsidiarity, to use one of the very many ugly words common in Union circles - but that often means going to a level below national boundaries, to region or city, or, as may become increasingly common, to regions which run across national boundaries. To devise a system, which places decision-making at the lowest possible level, enhances sovereignty. An insistence on the nation as the sole defender of sovereignty is old fashioned and unhelpful in the present fluid process towards European union.

One of the current problems facing the European Union, as it is at present constituted, is the need to progress on two fronts at the same time. On the one hand, the move towards greater union amongst existing members is in full flood - the deepening process. On the other hand, pressures to increase the number of members are also building up - the widening process. Apart from the enormous administrative task of trying to move on both fronts at the same time, there is potential conflict in the case of other European countries seeking to join the Union. New member countries may not be able to accept the full consequences of joining the Union immediately. In these cases transitional arrangements need to be agreed but the further the deepening process of the existing Union has gone, the more difficult transitional arrangements become. There is also a long list of other applicants, mainly from eastern Europe or countries previously part of the Soviet hegemony. In these cases, whilst their desire to join is just as intense, there is a much greater gap between the reality of their present political and economic institutions and those of the Union they wish to join. Considerable caution is thus being urged, even to the extent of suggesting that these countries, or some of them, may never be able to accept the full consequences of union; this would imply a two-tier system which could work but would be a pity. One of the results of the breaking up of the Soviet Union has been the realization that the centre of gravity of Europe, culturally at least, is far further to the east than those in the west have been allowed to admit over the past fifty years. If the European Union is to be truly European, it must fully embrace the eastern half of the region, despite the many and obvious difficulties in the way. The path towards European union has always been about challenges; the challenges of enlargement are as great and as exciting as the need to deepen.

As if the pot is not already bubbling with issues, the future Union will also have to consider the issue of where its boundaries finally lie. Ex-President Gorbachev in his excellent book [2] labels a section, 'Europe is Our Common Home' and by so doing lays a claim for Russia to be part of the eventual Union. This is a valid claim, for that part of Russia up to the Urals. But Russia's borders stretch across to Vladivostok. A Europe which stretches from the Atlantic to the Sea of Japan is hard to envisage.

Another difficult decision for the European Union is the candidature of Turkey; the first predominantly Muslim applicant for membership. Many criteria will need to be satisfied before the application can be accepted relating to conditions of democracy and methods of economic management. The most important criterion will be the relation between Turkey's value base and the common European heritage. It is not necessary to be negative. The Muslim tradition has always been part of Europe; in Spain and in eastern Europe. It is also a live tradition amongst immigrant groups in modern Europe. It does require some serious, but welcome, thought about the value base of Europe and what Turkey or other Muslim countries can bring to that debate. It is certainly indefensible to attempt to build a solely Christian Europe (assuming religious values continue to play a large role anyway).

Reference has already been made to the wider international consequences that might flow from the Union's intention to build upon communitarian values in its approach to economic and social policy. Such a policy is easily labelled protectionist and, in a sense, there is an affinity. Certainly the term preservationist would be appropriate, in a social and cultural sense. It implies that internal and world events will always be controlled with the objectives of regional social and cultural development in mind - a defence of regional sovereignty, some would say. Market capitalism has a massive impact on society, as has already been argued at length, whether resulting from internal forces or originating in other parts of the world. A healthy society and particularly a healthy European society, will not allow such forces free rein; they will need to be controlled in the interests of wider social considerations.

A market capitalist response to this view would be to claim that, by limiting free trade, Europe will prevent the world economy exploiting the comparative advantages of international trade. An answer to this argument was outlined in Chapter Five. Maximizing trade is not the sole object of economic policy. As Germany and Japan have shown, issues of growth in productive power come into consideration as well. Moreover, it is folly to assume that all countries subscribe to a belief in the merits of free trade. Free trade is the cry of the economically powerful (once the UK and now the USA). It is rarely an appropriate path for the underdog or the worldly wise. As Dudley Seers [3] has rather cynically argued, the conditions for successful development for a developing country are to have no valuable mineral resources (a big power will want control over them), a large population (it is less easy to be invaded then) and to be located as far away from a key power as possible (to avoid coming under its strategic defence interest). The point he makes is that it is these issues of real politik not the theoretical sketches of some economic academics that really matter. In any event, in the case of a big power, if a domestic industrial lobby feels threatened by external competition, it is sure to find ways of blocking the free flow of trade.

All this is not to say that the objectives of free trade should not be pursued. But they should be pursued realistically with wider considerations constantly in mind. Similarly, in the task of helping developing countries, whilst the challenge to help is an imperative, promoting free trade is not necessarily nor likely to be best policy to pursue.

The theme running through these remarks is that the prior question, for those living in the European region, is to answer the question 'what does it mean to be European?'. History does not give an unambiguous answer to that question. When people feel threatened, they often hark back to a past period of nationalist glory or take a (justifiable) pride in present national ways of doing things. The sense of national identity is there and it does not have to be denied or obliterated. It is to be found in France, Italy, Germany and elsewhere, as well as in Britain. However, there also exist strains in history that are common to most countries in the Union; where European can identify with European. It may be in music and other forms of art. It may be in a common religious heritage. It may be in a common democratic system of government. Or it may be in the shared experience of the industrial revolution. Or, indeed, in the common lot of war.

In all these trends, the national and the regional, different ideologies co-exist and make choice complicated. The over-riding conclusion to be drawn is that history is not going to make the choice obvious. As Edgar Morin argues [4], the decision for Europe is a question of will. Do people want to be European? Is there the will to be European? Do people want their foreseeable future to be mapped out in the context of this greater unity?

To express this challenge in this way perhaps appears to complicate matters. In reality it underlines that the issue facing the Union is to take a decision which shows that people are in control of their own destiny. People must decide who they want to be and where they want to go as individuals and as societies. The challenge is an emphatic rejection of those arguments which imply that people are victims of deterministic forces, of either history or the so-called iron laws of economics. People are free, free to choose, free to choose the appropriate many-faceted path towards human fulfilment. Europeans are being challenged to answer these questions within a new concept of Europe, which has not been considered, as comprehensively or as dynamically, since the Holy Roman Empire, a thousand years ago.

The wider view of economics expressed in this book is that economic activity, despite appearances to the contrary, is ultimately to be regarded as the servant of society. Economic activity is to be constrained to help achieve social targets and constrained to function to meet these targets within social conditions. The precise targets set by society, and society's views about the means by which these targets should be achieved, are matters of great debate. It may be obscure but society in one way or another is constantly engaging in that debate.

One of the tasks in that debate is to explore the extent to which those involved in the debate share common values. If there are common values, it is easier to move forward than if there is not. In the latter case it is the supreme task of the democratic process to find a way forward. Values are not set in stone forever. They move over time. Certainly the values people choose to pull off the shelf, so to speak, to match and support their inclinations, vary over time. Society is embarrassingly flexible in the way it responds to the challenge of changing technologies. It appears ready, each time, to adapt to ensure that the maximum exploitation of new technologies is made possible. Up to now that adaptation has been made mainly within the context of the modern state (with an international dimension given to it from time to time by the growth of empire). Now it is different. The present technological revolution is rapidly demolishing national boundaries and posing the challenge of how to globalize national institutions without total loss of control. President Reagan claimed (probably justly) that one reason why the Soviet empire collapsed was its inability to prevent new information technology crossing its boundaries in the form of satellite and space communications. The pervasive influence of international news networks, now to be found in almost every hotel room in the world, reflects the globalization brought about by information technology.

The impact caused by modern information technology on a global scale, and the inadequacy of national institutions to control it, is an example of the imperialism of technology and its wide ranging economic and social consequences. The way in which new technologies ally themselves with new ideologies and sweep virtually everything before it, has been examined at length. The defence of values, built up nationally, is everywhere being demolished by a global imperative in which technology demands free access across international boundaries. This pressure is producing a fatalism at national level. The individual sense of control over his or her future, which is largely bound up with his or her economic prospects, has been critically weakened. The same is true at national level where national governments are ineffectively seeking to combat or accommodate larger global forces.

Different conclusions are drawn from these trends. There is the argument that only exposure to global market capitalism will hone character to the point where people can survive. Countries/economies must therefore always be ready to respond to the next wave of competitive products from wherever it comes and at whatever cost. Even if it is desired to impose wider institutional constraints to ensure that economic activity serves society, national responses are increasingly inadequate. This is what makes the present great experiment towards European union so important. Things that can no longer be done by national institutions and for which adequate global institutions have yet to be created, can only be tackled at a regional level such as Europe.

The framework already existing in Europe is well suited to respond to this challenge. There is the legal framework in the Treaty of Rome, which adequately sets economic activity in the broader context of social objectives. The Union itself has always carried out its functions in the wider context of social goals. There is the evolving European parliament, which is becoming increasingly democratic, with more powers, which provides the first attempt at regional democracy. The emphasis, too, on the distribution of decision-making to the lowest possible area, will greatly help to counter-balance the growing disenchantment with remote transnational institutions. It will never be easy to govern a large region in a way that is adequately sensitive to local issues, but it will not be impossible. There is always the model of the United States to inspire.

People in Britain, of course, need to have their say in the way in which this Union is to work, especially in the choice of ideologies upon which to determine policy and set up institutions. It appears that at the moment Britain is out of step with the rest of the Union. This discord appears in many ways, but it is fundamentally reflected in two positions taken up by the present (1995) Conservative administration. The first concerns the issue of national sovereignty, where the administration feels issues of principle are at stake. But the position hardly makes sense unless it is seen as a proxy argument for the second key issue, upon which the administration takes an equally dogged position; that is over the ideology which it wishes to see underpin the market capitalist model of the Union. The threat to the British position, presented by the majority European view favouring a communitarian approach, is at the heart of the matter. The defence of national sovereignty is a convenient way of trying to ensure that Britain does not have to adopt what it regards as an alien approach to economic activity in particular.

It is hard to agree with the present administration's stance on these two positions (indeed, it is probably only one position). There is as much in history to encourage Britain to go with its partners as there is against. Also it seems essential that Britain works with this regional approach to controlling Union affairs, social as well as economic, otherwise Britain will be blown in all manner of directions by the impact of global technology.

Present moves towards European union are the supreme test of the view that a healthy society (at no matter what level) requires a means of controlling the way it chooses its objectives and the means of achieving them. It also requires a means of ensuring that economic activity is kept in its place so that it will not run rough-shod over other aspects of society or culture; rather, that it will be run in such a way as to contribute to life fulfilment as perceived by the majority of citizens.

If this is to happen effectively, European society needs to ensure that a proper debate is conducted about the value sets on which to base social and economic activity. The various partners in the union, including the British, have to decide with whom they share culture - which emphatically points Britain towards Europe rather than towards, say, the United States. As a community, the Union has also to decide what price it is prepared to pay to defend its culture. In view of the over-riding importance of wider social activity in life fulfilment that price cannot be too high.

Adequate means of discussing and identifying the Union's ultimate objectives are also needed. That does not mean a clearly defined process seeking clearly defined ends. Life is never that simple. Open debate is important, however inconclusive or imprecise, and society must work to achieve that - it certainly cannot be left solely to the politicians. It does not mean setting up a vast bureaucratic structure to make it all happen. Besides the impracticality of such as venture, the process would be more than likely stifling and self-defeating. But open debate about the means people wish to use is crucial.

Finally, somewhere in this social process, room must be found for the voice of men and women of goodwill to be heard, as Claude Gruson would put it; that voice is ethical and is therefore designed to encourage society to higher things. It would also, and society should be ready to listen, wish to judge where necessary. Society needs its builders. It also needs its prophets. These are the people that can encourage by pointing out true objectives whilst disturbing complacency by underlining the gap between what is and what could be.

Returning to the original question of what is the purpose of the move towards a greater European union. The objectives undoubtedly include the economic goal of greater efficiency and competitiveness. The record of the Union, so far, shows the extent to which that has been achieved. Further gains in efficiency, to improve performance around existing technologies and to introduce and exploit new ones, are desired and there is no reason not to expect them to be delivered, given past records. These improvements in efficiency will allow the Union to continue as an effective player in the world market.

However, the over-riding objective of the Union must continue to be the pursuit of the means whereby everything, especially economic activity, can be harnessed effectively to serve European social goals, in a way consistent with the view European men and women hold of social life fulfilment. Society (individual and community activity) must reign supreme and it must be true to its values.

Notes

1. Seers, Dudley (1983) makes this and many allied points effectively in his book, *The Political Economy of Nationalism*, Oxford University Press
2. Gorbachev, Mikhail (1988), *Perestroika*, Fontana/Collins, London, p. 194
3. Seers, Dudley, op. cit.
4. Morin, Edgar (1987), *Penser l'Europe*, Editions Gallimard, Saint-Amand (Cher), France

13 Epilogue: The unblessed venture

The over-riding feature of western society is its materialistic base. Some two centuries of market capitalism has resulted in a massive growth in the capacity to produce goods and services. The appetite to consume appears to have grown at least as fast, to the point where the acquisition and use of consumer goods and services is central to much of social activity. The sheer size of the national product, the amount of activity devoted to sustaining or increasing it generated, still, by a perceived shortage of resources for meeting individual and social choices, sets our society aside from any other.

Given the importance of this materialistic base in our daily lives, it is right for society to be concerned about that part with the responsibility for producing goods and services; the economic model, so to speak. Society needs a system which allows choice to be expressed and which ensures that resources are used efficiently, with no waste. Few would disagree with the view that market capitalism meets these criteria. Market capitalism in its present state is also robust enough to underpin today's complicated inter-related global economic system. There is no effective radical alternative. Most debate about alternatives revolves around modifications to this market capitalist model.

Debate about control of the economy has to be set in the context of the dynamic nature of economic activity. The past reveals a continuous process invariably beginning with technological innovation leading to new forms of processes and new products. Most of the time this pattern of change is revolutionary with dramatic repercussions. In narrow techno-economic terms the process is well summed up in Schumpeter's phrase 'creative destruction', where new drives out the old. The process is complicated and invariably takes time but it underlines the key feature of economic activity - an atmosphere of constant change. Change is mainly about power, the creation of productive capacity and the struggle between systems based on different technologies fighting for supremacy or survival. It is no surprise that the debate as to who has control over, or responsibility for, guiding this dynamic system takes place in the political arena.

The modern economic system both creates and reflects a number of features of market capitalism. It dictates a particular lifestyle for people in society. As producers, people appear to be locked into, willingly or not, an unremitting work effort in the name of a constant fight against competition. As consumers, people appear to be prisoners of an almost frenetic pattern of consumer spending which they see as essential to their attempts to satisfy immediate personal and social goals. People find themselves, as producers or consumers, on a treadmill which constrains most of their daily lives. Consumerism has been too successful; it has turned master into servant and servant into master.

Changes initiated by the economic system have invariably resulted in large scale social dislocation; such as the consequences of a factory form of production in the nineteenth century or the disappearance of that system in the late twentieth century. The past shows the radical nature of the impact on social institutions and lifestyle; equal in power and effort as the changing economic factors themselves. Most consequences were not foreseen but it is possible that some of the changes had an element of intent in them - changes brought about to remove perceived social obstacles to the full exploitation of new techno-economic forces. The need to adjust applied to value sets as well.

Although society at large provides a superior structure to the limiting one of economic activity, it (society) has a constant struggle to maintain its supremacy. It invariably finds itself in a reactive role and is constantly put on the defensive. In so far as existing social structures are attacked by new economic forces, a form of creative destruction also takes place in the social sphere. Institutions may be found wanting in coping with change and therefore become obstacles to change. They may resist, but normally will be overcome eventually. Institutions may be attacked in this way; so may the value sets underpinning them. These institutions can aptly be called heavy structures and they are constantly exposed to change. The apparently inevitable pressure is not always to be yielded to. In some cases the heavy structures contain a legitimate interest that need to be defended. In other cases the defence is inappropriate and they should make way for the new.

Nonetheless, it is right and proper to look to the wider social context for factors to be put in the balance against the forces of economic change. These factors should embody, first, an expression of how society wants to control and organize economic activity and, second, an expression of how society wants to use the fruits of economic activity.

Some of the factors society invokes to control and organize economic activity may take the form of constraints. They may relate to working conditions, pollution control, income distribution (setting maximum or minimum wages), monopolies legislation etc. More positively, wider social values may suggest better ways under which business activity can be carried on; by encouraging forms of management practices in tune with broader social norms or encourage different forms of corporate social responsibility. These latter factors are far from being constraints. They are the basis for building successful business activity. Despite impressions to the contrary, models of management and forms of business activity which have been successful, are not uniform either in comparison with the past or across today's world.

However, the most important social task is to ensure that society has in place the means by which decisions about the use of the fruits of economic activity can be made. Such choice will be individual in, for instance, the use of the market for individual consumer choice. The decisions will, however, also be social. Whichever, society must ensure that an effective ethical debate over means and goals is carried out. It is not an easy task but if it succeeds it will demonstrate that society, not the economic system, reigns supreme.

To quote Joan Robinson again, 'Any economic system requires a set of rules, an ideology to justify them, and a conscience in the individual which makes him strive to carry out this end'. [1] No-one would deny that these features apply to market capitalism. There is a vast body of rules designed to explain how the economic system works. Thankfully, at last, the underpinning ideology has been brought out into the open rather than hidden within the rules. Also individuals have been encouraged, in true libertarian fashion, to work within the system to satisfy self and conscience.

It would be idle to seek to argue that the libertarian view of behaviour should be eliminated as a source of motivation in economic activity. It would be equally idle to try and remove community values entirely. There is a place for both. The argument is about balance. As far as consumer behaviour is concerned it seems desirable to acknowledge the value of the market in providing a means for allowing individual choice to be expressed (subject to the qualification that sociologists would want to place on the role of individual as distinct from social behaviour). Whether an individualistic view is as appropriate as a basis for management models is a matter for more dispute. Cooperative management models have proven to be very effective. There is much evidence also to suggest that such models are more in tune with British culture than a more individualistic style.

Joan Robinson's remark can be applied equally to social as well as economic systems. There are other ideologies underpinning other institutions or reflected in common law and customs and in attitudes to welfare; the latter being the locus of a fierce ideological debate between libertarian and community values. It would be helpful if the task was simply to find ways by which different ideologies could co-exist in society. Unfortunately, peaceful co-existence is not the norm. Ideologies, particularly new ones associated with change, can run rampant and seek to spread into other areas. As there may be good or bad reasons for this imperialism, mechanisms are needed to enable the pressures for change to be openly debated and judged. The inevitability of uncontrolled change has to be challenged and an ethical dimension permitting control allowed to come in. It must be possible to be able to ask the question 'Is it fair?' in every case - whatever the ultimate answer to the question turns out to be. How society answers the question will depend on circumstances, the prevailing balance of ideologies and the constraints of culture and tradition.

It is important to open up the debate over the relationship between economic and social (cultural) activity. The cultural background of current economic models in other regions of the world are relevant. The Japanese approach where the company is seen as a community, the importance given to planning and the country's present 'latter-day mercantilism' as a means of furthering Japanese interests, underline two important points. The first is that there do exist modern economies which do not wholly conform to the universalist view of economics, designed to optimize trade rather than enhance economic power. The second is that Japan is an example of a way in which distinctive non-libertarian cultural values are incorporated into an economic model with impressive consequences.

Whilst not so distinctive, the European tradition is similar in that it has never entirely trusted models of unbridled market capitalism. That tradition also contains a close link between business and the state and an acceptance of the role of the state in community to constrain and guide the economy as well as support society. Much of this tradition stems from the impact of Bismarck and Napoleon, but British history contains many similarities. The tradition also contains the concept of social partnership, in which the respective objectives of good labour relations (including wage determination), successful business performance and a satisfactory wider social welfare system, are jointly determined. Here, too, is a practice some distance away from pure libertarian tradition.

In the case of the United States of America there are identifiable links between culture and ways of doing business but in this case the model is more consistent with a libertarian ideology. On the other hand, the United States has its communitarian and interventionist traditions.

The message for Britain from these examples is that it must address the British situation in the context of British culture. That is the grounds for discussing economic success; not narrow economic parameters, such as wage costs or production. The challenge is thus more about education and class, about cultural objectives and appropriate management models. It has to be reluctantly accepted that Britain is not clear as to the cultural base that it wishes to take into the twenty-first century. The traditions of empire are still too close and possible links with the United States ever tempting for it yet to take a liberated view. But eventually it must return to its European homeland, its European culture and pursue ways of doing business consistent with its partners in the region.

The present lines of development in the European Union have much to offer Britain. The sheer size of the market of the region will put sufficient competitive pressures on Britain's economy to ensure its efficiency. Also, the parameters being set by the Union to secure its future, offer Britain an exciting opportunity to build on the richer seam of its own historical strengths. These parameters are mainly social and cultural, many of which are incorporated in the principles underlining the social chapter of the Maastricht Treaty. The majority of Union members take the legitimate view that it is appropriate to set standards for, for example, normal working hours, safety at work, the treatment of women and part-time workers in employment as well as standards for the purchase and sale of goods and services. Not to take these social and cultural parameters into account reflects a seriously diminished view of the range of objectives that a market capitalist economy should set itself.

Economics is not universal. The way countries perform depends on geography, the availability of resources, their particular cultural attributes and the timing of their development. The particular social context of a society dictates the ideologies upon which a successful economic model should be built. It is necessary, therefore, to ensure that economic activity is carried out in a way that does not damage that social context. The practice of market capitalism must therefore always be conditional in this cultural sense. There is an immense power of performance unleashed when cultural values are properly harnessed into the economic model. The USA, Japan and Germany all have highly successful economies, but each has significantly different ways of going about things. The most important of these differences reflects cultural differences. So societies have to choose the economic model that best suits their culture.

One of the purposes, perhaps the key purpose, of this book has been to establish that ultimately economic activity has an ethical dimension to it. That dimension is reflected in the fact that the results of economic activity - the flow of goods and services - is one of the principal contributions towards life fulfilment in western materialistic society. It is also reflected in the means by which economic activity is organized. That activity concerns men and women working together and there are good and bad ways of organizing people. That choice, of how to work together, also contributes to life fulfilment. Economic activity is also ethical in that, in order to succeed, it needs to be based on the cultural values to be found within society at large. Conformity with these values will ensure that the processes and the distribution of the fruits of economic activity are fair. Those responsible for economic activity cannot avoid being judged ethically.

If economic activity is to be elevated onto an ethical level in this way, it is necessary to offer some guidelines by which the system can be judged; in other words to give meaning to what is meant by a good or healthy economy. There are many indicators of health. One is that the economy is run efficiently. This must be the prior judgement. It is what all economists feel they are in business to provide; a system that produces goods and services with a minimum of inputs. Where resources are allocated efficiently to meet given ends. No system is perfect, but most western economists believe that the market place is the best system so far devised for the exercise of choice, for allocating resources for ensuring the optimum use of goods and services.

Second, in addition to being efficient, an economy has to be effective; i.e. it has to achieve its objectives. A major task for any society is to decide what to do with resources created year by year by the economic system. The processes used by society to establish objectives are complicated and often obscure. They include straightforward economic choices by individuals based on purchasing power, but they also include other motivations which many sociologists have sought to identify. However, the question of how to set aims and objectives generally remains. Some, however, might say that people are not effectively permitted to have aims in a materialistic society dominated by technology and by advertising. Many spending patterns appear to be aimless, but we can take comfort from the fact that, at the heart of the economic model, is the consumer. He or she really needs to be made sovereign (in the sense of having control over his or her life fulfilment, where that goal is pursued by the purchase and use of goods and services), if there is to be any hope that society can be master of its own fate.

What are the real purposes of consumption? A number of references have been made to the views of Mary Douglas and others that the reason for consuming goods and services is to do with more than achieving individual satisfaction. There is a social dimension to the use of goods. Goods are given value by others. They are used to demonstrate a sense of worth and to indicate who people want to be seen to be. Goods are used as markers and as a means of social communication. Social rather than individual decisions are paramount.

> 'Consumption is the very arena in which culture is fought over and licked into shape. The housewife with her shopping basket arrives home: some things in it she reserves for her household, some for the father, some for the children; others are destined for the special delectation of guests. Whom she invites into her house, what parts of the house she makes available to outsiders, how often, what she offers them for music, food, drink, and conversation, these choices express and generate culture in its general sense.' [2]

> 'Consumption patterns ... are ultimately moral judgments about what a man is, what a woman is, how a man ought to treat his aged parents, how much start in life he ought to give his sons and daughters; how he himself should grow old gracefully or disgracefully, and so on.' [3]

Time and time again the argument has come back to the conflicting stances of the libertarian and communitarian approach both in relation to economic activity and to the wider task of life fulfilment in society. In principle, it should be possible to describe legitimate applications of each ideology which could be accepted and allowed to co-exist. In practice, the competition for dominion between ideologies appears to be too severe for such an easy accommodation. The defence of libertarian/individualistic virtues as a basis for certain parts of economic and social activity is well made; even though the distinction between self-interest and selfishness is often lost. The defence of communitarian values appears to be harder, although for those who approach the task from a religious point of view, it would be argued that this is not because the case is weaker. Far from it, the case for looking beyond the immediate needs of self and seeing and responding to the needs of others brings in a wider moral dimension.

Sharing is perhaps the best word to encapsulate much of this extended moral dimension. One of the immediate applications of this morally loaded term is to the process of sharing rewards. A key question in economics is to settle rival claims to the surplus created by economic activity; and the associated claim to the right to property. As far as income distribution is concerned market capitalism provides its own way of establishing the rewards to factors of production. They are determined by the balance of supply and demand in the market place. It must also be remembered that market capitalism revolves around risk-taking entrepreneurs and the role of enterprise activity generally. Society ignores the importance of this role at its peril. It is not simply a question of moneymaking. It is to provide an environment in which such people, all people for that matter, can fulfil their fundamental ambition to be stretched to their limit. Associated with this activity is a heroic view of humanity, of the practice of positive and aggressive virtues. It is the view that life is a gamble where there are great gains to success, but with a high risk of failure. The economic success of western societies owes a lot to this view and equity demands that these attributes should be rewarded.

However, a measure of the quality of a society is its ability to share income - irrespective of merit. The willingness to accept that this means active involvement in society, accepting some income redistribution through taxes, to support the victims of economic change, is currently weakening in Britain. However, it is a crucial area for practising justice. Oddly enough the problem appears to get no easier as societies become richer. A withdrawal into self-interest is understandable in a society living at survival levels. It would seem natural to assume that, as a society draws away from survival conditions, an increase in the ability to share would be matched by an increase in the willingness to share.

There is also the vexed question of the ownership of property. Possession of property is a vital motivator in economics. However, most religious teaching has underlined the conditional nature of ownership and the need to practise the concept of stewardship and responsibility to others. This point is clearly made in Pope John-Paul II's Encyclical on Human Work. The Encyclical's view of the role of property is claimed to be different 'from the programme of capitalism practised by liberalism and by the political systems inspired by it'.

> '... the difference consists in the way the right to ownership of property is understood. Christian tradition has never held this right as absolute and untouchable. On the contrary, it has always understood this right within the broader context of the right common to all to use the goods of the whole of creation; the right to private property is subordinated to the right to common use, to the fact that goods are meant for everyone.' [4]

More pertinently and personally, all of us need to be humble about our own property; our education, the luck of our birth, parents, country, our genes, our luck in business. This property is not owned by us. It is lent to us to use responsibly.

Another area for practising sharing is in the way in which people work together in community. People need to share in the process of economic activity. This is particularly true in times when people are experiencing rapid change; a sense of cohesion is essential. People have to feel good to work well, which they will not; if they feel threatened by the move from the goods to the services sector; if they are threatened by the need to move from big companies to individual networking or from a hierarchical to a flat management structure, or from a work dominated to a leisure dominated society, or from working in one area to working in another. All these elements of change can be threatening in the extreme. Understanding and cooperation are required to turn them into life-fulfilling challenges and opportunities.

There is a crucial role for good management models both in business and as an element in social activity. Whatever their form, they need to reflect two key features. First, they need to allow people to work in a way that, at the end of the day, they are able to go home satisfied. For most, they will do so if they have been paid properly for their effort. But, above all, they need a deeper satisfaction relating to team effort. They need to be able to say, 'that was a good day and what we achieved, we achieved as a team'. Second, much revolves around the role of leadership in management. Leaders need to have authority, but they cannot claim authority; it has to be given to them by those they seek to lead. Therein lie the elements of a good and effective participatory management model.

Most people would recognise that there are categorical imperatives to sharing. Those who deserve our help include those living in the Third World - an area of challenge to the west about which no more will be said. But there is a similar moral imperative attached to the challenge of sharing with the weak in our own society. Mary Douglas makes the point that in a primitive society, practices of exclusion, which underline the distinction between rich and poor, can clearly be seen as a result of the way in which society organizes itself. But in western societies the distinction between rich and poor is not regarded as an outcome of similar social behaviour. Instead poverty is seen to be a result of the malfunctioning of the economic system. As she goes on to say,

> As a result, the problem of poverty in the midst of industrial plenty is seen solely as an outcome of the system of production to be solved by redistributive legislation and state control. [There is a] complementary view. The poor are our kith and kin. Not all our relatives are likely to be among the well-to-do. If we do not know how the poor live, it can only be that we have selected against them in the constituting of our own consumption rituals. [5]

This is the ultimate moral challenge created by economic activity providing current levels of affluence; to close the gap between rich and poor, between the House of Have and the House of Want, as Henry George put it over a hundred years ago. Society will hopefully accept the challenge to eliminate this distinction as best it can. But until it does each state, of riches and poverty, needs to be judged.

> Wealth is not a god to be worshipped; nor an idol to be shattered. It needs a context. It needs religion ... Poverty ... is an evil of theological dimensions. But if poverty is unconditionally evil, affluence is only conditionally good. It requires that wealth, work and the human world are pursued to religious ends. [6]

A few years ago when the British Broadcasting Corporation celebrated its fiftieth anniversary, a committee was formed to prepare material to put in a capsule to bury for future posterity. One of the participating members was a scientist called Ritchie Calder, and whilst I cannot recall or trace his exact words, his contribution was in the form of a very simple and short letter which went something as follows:

Dear citizen of the future,

I have no idea what the state of the world will be when you open this letter. It may be that you are living in a land of immense plenty and enjoyment, or you may find yourself in a world devastated by disaster.

If it is the former, I simply wish you well. If it is the latter, I can only apologise on behalf of our society. We have worked on the twin assumptions that all technological change is progress and that, on balance, man is good. I have to say that neither of these statements is certain.

I have said much in this book about technology and its impact on economic and social structures. The open frontiers still presented by the flow of new technologies remain enormous. It is reasonable to assume that the next hundred years will contain as much technological change as the past century. It is reasonable to assume, also, that such changes will have the same radical impact on the ways of doing economics, on the composition of the fruits of the system; with similar radical consequential social and institutional changes. The ideologies underpinning the present system will no doubt be as severely challenged by change as past value sets have been. Change is not the same as progress. Technologists are tempted to assume that there is a technological solution for every problem or, more ominously, that there is a good use for each new technology. Society should be cautious; ever ready to limit unimpeded exploitation of new ranges of technology.

Society still needs to be on its guard even where the introduction of new technology produces broadly acceptable change and innovation; given the scale of economic and social consequences likely to occur. Society needs to take an even more guarded views of certain areas of technological experiment where the risks of adverse effects are high. The environmental impacts of current technologies, of which the motor car is the blatant example, or of energy supply systems causing acid rain, or raising fears of nuclear safety have rightly attracted the public attention imposing social constraints on further developments. The challenges of new technologies are even greater, especially in the field of genetic manipulation intruding on what hitherto had been regarded as forbidden territory for scientific and technological development.

As regards the issue of human nature, the story is also ambiguous. No oversight of the way in which the economy in society works will succeed unless the nature of human appetite is recognized. Human beings appear to be essentially selfish. Most attempts at social engineering, 'to build a better world', have either led to failure or to alternative forms of tyranny. On the other hand, it would be a very pessimistic society indeed that assumed that there was no element of altruism in human nature to act as a basis of a welfare society. The dilemma is that if the degree of altruism is assumed to be greater than it is, all attempts at social engineering will come to nought. The consequences of assuming that there is no altruism are not worth contemplating.

This book has been primarily about economics and its role in society. However, if we are to be satisfied with that role, all things need to be focused on the human being - the precious heart of it all. As Ruskin was alleged to have said, 'The only wealth is life'. It follows, therefore, that in all constructions of models, both economic and social, there is an over-riding need to protect society against tyranny and, in so doing, encourage individuals to flourish, as individuals, since all count, and within a broader social context in which to develop the concept of person able and willing to look beyond narrow self-interest.

In one of R.S. Thomas' poems [7] he talks about the human spirit (*The Hand*) wanting to change the world and seeking God's blessing for its activities. But, 'God, feeling the nails in His side' said, 'I let you go but without blessing'.

As an economist I reflect much on the genie that was let out of the bottle at the end of the eighteenth century. The mix of human endeavour, reflected in technological and economic change, and the moral imperialism associated with that change, has pushed everything aside as it has swept through society over the years - and still continues to do so. The vast potential for good in that expression of human restlessness is in marked contrast to our lack of control and our inability to direct it. Can we find a way of gaining God's blessing?

Notes

1. Robinson, Joan, op. cit. p. 18
2. Douglas, Mary, op. cit. p. 57
3. Douglas, Mary, op. cit. p. 58
4. *Laborem Exercens*, Encyclical letter of the supreme Pontiff John-Paul II on human work, Catholic Truth Society, publishers to the Holy See, London, 1981 pp. 50-51
5. Douglas, Mary and Isherwood, Baron, op. cit. p. 204
6. Sacks, Rabbi Jonathan, informal talk given at St. George's House, Windsor Castle, November 1978.
7. Thomas, R.S. (1993), *Collected Poems, 1945-1990*, J.M. Dent, London

Index